Myofascial Manipulation

Theory and Clinical Application

Second Edition

Robert I. Cantu, MMSc, PT, MTC
Group Director
Physiotherapy Associates
Atlanta, Georgia
Adjunct Instructor
University of St. Augustine for
Health Sciences
St. Augustine, Florida

Alan J. Grodin, PT, MTC
Regional Director
Physiotherapy Associates
Atlanta, Georgia
Adjunct Instructor
University of St. Augustine for
Health Sciences
St. Augustine, Florida

8700 Shoal Creek Boulevard
Austin, Texas 78757-6897
www.proedinc.com

pro·ed
An International Publisher

© 2001 by PRO-ED, Inc.
8700 Shoal Creek Boulevard
Austin, Texas 78757-6897
800/897-3202   Fax 800/397-7633
www.proedinc.com

The author has made every effort to ensure the accuracy of the information herein. However, appropriate information sources should be consulted, especially for new or unfamiliar procedures. It is the responsibility of every practitioner to evaluate the appropriateness of a particular opinion in the context of actual clinical situations and with due considerations to new developments. The author, editors and the publisher cannot be held responsible for any typographical or other errors found in this book.

Library of Congress Cataloging-in-Publication Data

Cantu, Robert I.
   Myofascial manipulation : theory and clinical application /
  Robert I. Cantu, Alan J. Grodin. -- 2nd. ed.
      p. cm.
   Originally published: Gaithersburg, MD : Aspen Publishers,
   2001.
   ISBN 0-944480-67-5 (alk. paper)
   1. Myofascial pain syndromes--Physical therapy.
   2. Manipulation (Therapeutics) I. Grodin, Alan J. II. Title.

   RC925.5.C26 2006
   616.7'4--dc22

                                             2006045221

Printed in the United States of America

To my dear wife
Ruth
for her years of support, expressions of confidence,
and for helping me keep it all in perspective,

and to my son
Samuel
for his enthusiasm, zeal for life, and for
helping me keep the spring in my step.

R.I.C.

To my wife
Carol

and my children
Jason, Evan, Seth, and Lisa
for their support and tolerance
of my personal and professional life.

A.J.G.

Table of Contents

Contributors

Deborah Cobb, MS, PT
Physical Therapist
Physiotherapy Associates
Atlanta, Georgia

Jan Dommerholt, MPS, PT
Director of Rehabilitation Services
Pain and Rehabilitation Medicine
Bethesda, Maryland
Vice President
The International Myofascial Pain Academy
Schaffhausen, Switzerland

Clayton D. Gable, PhD, PT
Assistant Professor
Department of Physical Therapy
The University of Texas Health Science Center
  at San Antonio
San Antonio, Texas

Preface to Second Edition

When we published the first edition of *Myofascial Manipulation* in 1992, we were not fully aware of the interest and pent-up demand for this material. Since 1992, the book has continued to sell copies, and this has been a humbling experience for us. We believe there are several reasons for the continued interest in this material.

First, an underlying philosophy and strategy for the book was to provide good "bread and butter" techniques that were effective on patients, were relatively easy to learn, and were practical to use in the current arena of managed care. For the second edition, we have added a number of other "bread and butter" techniques, being careful not to add any "fluff" to merely make the book bigger. What are still represented in this edition are the myofascial techniques that the authors have used successfully over the years on a daily basis on literally thousands of patients.

Second, the first edition relied heavily on basic science principles. We went to the literature, for example, to explain the mechanisms of injury and repair, and to delineate pain of mechanical versus nonmechanical origin. We carefully extrapolated and integrated these principles into the principles of management and treatment of soft tissue dysfunction. For the second edition, we wanted to strengthen that scientific foundation. To that end, we enlisted the help of gifted professionals and content experts, to add

material, and to re-tool and revise existing material in the previous edition. The chapter on neuromechanical aspects of myofascial pathology and manipulation, for example, adds a dimension of understanding we did not offer before. Also, the chapter on muscle pain syndromes (i.e., pain of mostly nonmechanical origin) was completely rewritten due to the explosion of research in that area. The chapter on the histopathology of connective tissue has also been completely updated due to advances in research over the last 8 years.

As we mentioned in the first edition, *Myofascial Manipulation* is not designed to be a panacea for manual therapy, but a great utility tool to be used in conjunction with joint mobilization and exercise. In our courses, we often refer to that triad (soft tissue mobilization, joint mobilization, and exercise) as the "pinball triad of manual therapy." This is because the three aspects of treatment are virtually inseparable and totally integrated in the clinic. The savvy clinician knows how to effectively "bounce off" all three aspects of treatment to arrive at the desired, optimal result.

We respectfully submit the second edition of *Myofascial Manipulation* for your consideration as a tool to help expand the horizons of our profession. Managed care, Medicare cutbacks, market saturation of therapists, and turf erosion have put us in a position where it is no longer an option for us to be the very best. Our professional lives and the health and longevity of our profession in general depend on it. We hope that this tool will be useful in helping us all forge ahead to expand our individual and collective horizons.

Robert I. Cantu
Alan J. Grodin

Preface to First Edition

In his classic book, *Joint Pain*, John Mennell wrote that "no textbook in the field of orthopedics can be entirely original." On first reflection, this statement seems a bit contradictory, in light of the fact that Mennell was quite an innovator and one of the early advocates of using arthrokinematic rules for joint mobilization. On further reflection, however, his ideas and philosophies, while quite innovative, were based on a combination of knowledge and clinical experience he attained throughout his years as a medical student and as a physician. The knowledge and experience he gained over the years were molded and integrated in a way that became uniquely his own. His system became his "handwriting," or his style.

Handwriting is a good analogy for personal style. A person's handwriting is a totally unique self-expression. The uniqueness comes from the actual process of learning how to write, from years of practicing that handwriting, and from the particular function the handwriting serves in the person's life. A physician who has taken voluminous notes throughout school primarily for his or her own benefit will have very different handwriting from the architect who has to submit drawings with very legible writing. The letters formed in the handwriting, as well as the spelling, are not unique, but the way the letters are represented by the individual are.

So it is with this book on myofascial manipulation. For us, it is a combination of acquired knowledge and clinical experience that, over the years of treating patients, has evolved into a particular philosophy or system that is unique. For anyone to say that they were the first in history to "invent" certain techniques would be presumptuous. What we attempt to do in this book is to take the most current body of research in myofascia and integrate this cognitive knowledge with psychomotor skill to produce a concrete system of evaluation and treatment acceptable to a profession that is striving for higher professional recognition.

This textbook is divided into three parts that reflect its major purposes. The first part outlines the evolution of myofascial manipulation, incorporating both its history and the latest schools of thought. The second part and purpose of this textbook outlines the scientific basis of myofascial manipulation. Management of certain clinical problems is also discussed. Part III focuses on evaluation and treatment techniques that have repeatedly proved effective in the clinical setting and includes an atlas of therapeutic techniques.

For the sake of clarity throughout the text, manual therapy is divided into joint manipulation and soft tissue manipulation. As understanding of connective tissue has increased, the distinction between joint and soft tissue manipulation has become somewhat clouded. Joint manipulation has been defined as "the skilled passive movement of a joint." The tissues being mobilized, however, are all histologically classified as connective tissues, and in this respect, any type of manual therapy can be considered soft tissue manipulation. The distinction made in this text is in the arthrokinematics, or lack of arthrokinematics, in the application of the techniques. Soft tissue manipulation is generally less concerned with arthrokinematic rules than is joint manipulation; a majority of the techniques are not concerned with individual joints but with myofascial relationships and the interrelations of the joints to the soft tissues. For the purposes of this text, we have defined *myofascial manipulation* as: The forceful passive movement of the musculofascial elements through its restrictive direction(s), beginning with its most superficial layers and progressing into depth while taking into account its relationship to the joints concerned.

Myofascial Manipulation is not meant to be a panacea or an exhaustive critical review of the literature, but a representation of what we feel strongly about clinically. These are techniques that we use every day, integrating them with joint mobilization, alternate somatic therapies, and exercise. Our hope is that this information will be integrated into the readers' arsenal of techniques and into their philosophy of treatment, so that each clinician's style, or "handwriting," will become more distinct as well as more effective.

Robert I. Cantu
Alan J. Grodin

Acknowledgments

The authors thank the following persons for their assistance in the preparation of this volume: To Trevor Roman for shooting the photos in Chapter 8, and to Debbie Cobb and Brad Foresythe for being the "therapist and patient" in Chapter 8.

The authors also acknowledge all the professors who adopted the first edition for their courses and curriculums—the long-term success of this book is due to your support and votes of confidence. Thank you.

From the First Edition

The authors thank the following people for their invaluable assistance in the production of this book: Karen Barefield, PT, for her drawings in Chapters 6 and 7; Paula Gould for her photography in Chapters 6 and 7; Carolyn Law, MPT, for her help in editing the manuscript, both from a content and grammatical standpoint; and Lisa Richardson, for being the "patient" in Chapters 6 and 7.

PART I

Historical Development and Current Theories of Myofascial Manipulation

CHAPTER 1

Historical Basis for Myofascial Manipulation

Robert I. Cantu

Myofascial manipulation is as old as history itself—humans have been performing myofascial manipulation as long as humans have been touching. Throughout history, many different systems and supporting theories for the treatment of musculoskeletal pain and dysfunction have come and gone. Today, the originality of any current system of manual medicine is generally found in the underlying philosophy, not in the techniques themselves. The underlying theory and philosophy of any manual therapy system will dictate the sequencing of technique, and will attempt to explain both the results and the proposed mechanisms of action. The techniques may be old, but the packaging is new. Underlying theories may alter the way the treatment is performed and may vary and modify the technique. The advent of the scientific age has yielded a tremendous wealth of scientific information, which in turn has changed the theory and philosophy of modern manual medicine.

Currently, and throughout history, the scientific thinking of the day has fashioned the existing schools of thought in manual medicine. We treat based on what we know or think we know. The purpose of this chapter is to chart briefly the evolution of manual therapy, with an emphasis upon myofascial manipulation. As the different historical trends are addressed, a greater appreciation of current manual therapy will be gained.

The evolution and persistence of manual medicine throughout the years have been remarkable, especially since the medical communities often shunned such treatment, and its scientific basis has only been heavily researched within the last 40 to 50 years. This research has fostered a redefinition of manual medicine and a redefining of exactly what is being accomplished with manual therapy.

The history of manual medicine can be divided into four basic time periods. The first period, which begins in ancient history and ends roughly at the close of the nineteenth century, emphasized *position*. Joint pain, including spinal pain, was a result of a "luxation or subluxation" of one or more of the joints. The emphasis in the spine was in restoring the position of the vertebra to relieve pain. In the second time period, starting with the early twentieth century, the philosophy and theory of manual medicine began to emphasize *mobility*. Restoring mobility to a joint that "was locked" became the focus of manual medicine. The science of arthrokinematics developed, and terms such as "accessory movements" appeared. This spurred the curiosity of researchers in the mid and late-twentieth century, who pushed the study of manual medicine into a third phase—understanding how manual therapy affects the *biomechanics of connective tissue*. They viewed the increased mobility of the joints as a result of mechanical changes in connective tissues. Largely because of the chronicity and recurrence of many types of back pain, the present period of research in manual medicine is beginning to concentrate on *neural mechanisms* of back pain and *movement reeducation* (see Chapter 5 for discussion of neural mechanisms in myofascial therapy). The science of motor learning and control will have much to offer in this area. The immediate future of manual therapy lies in the combination of passive manual therapy techniques and movement reeducation or motor learning techniques for prophylaxis. Each of the different time periods and their underlying philosophies is discussed in the following sections.

ANCIENT TIMES

Early recordings of manual medicine date back to the time of Hippocrates around the year 400 BC. Two relevant works, which include "On the Joints," and "On Setting the Joints by Leverage," describe various combinations of manipulations, massage, and traction on a wooden table.[1] Much of Hippocrates' work in early manual medicine can probably be attributed to the popularity of wrestling in his day. Entries in early manuscripts include descriptions of both joint manipulation and massage in treatment of a dislocated shoulder.

> The next patient is a sleek Ephibos, still oily from his last wrestling match in the gymnasium. He is clutching his left arm, obviously dislocated at the shoulder; the pain is not great, and it is the fourth time it has happened, anyway. The treatment was routine to him.... The main problem is solved once again; and if the maneuver has failed, the gladiator had eight other ways to go about it, by pulling the patient's arm over the chair.... And it is necessary to rub the shoulder gently and smoothly. The physician must be experienced in many things, but assuredly also in rubbing; for things that

have the same name have not the same effects. For rubbing can bind a joint which is too loose and loosen a joint that is too hard. However, a shoulder in the condition described should be rubbed with soft hands and, above all things, gently; but the joint should be moved about, not violently but so far as it can be done without producing pain.[2,3]

In the treatment of back pain, Hippocrates describes treatment of humpback, or alternately translated "kyphosis." Hippocrates is probably referring to a kyphosis of the lumbar spine. He describes two treatments for this condition consisting of mechanical traction and extension exercises.

> If possible, the patient is first given a steam bath...then he is placed on his stomach on a wooden board [for traction].... The physician places the flat of one of his hands on the kyphosed portion of the patient's back, and his other hand on the top of the first.... He presses vertically, or in the direction of the head, or in the direction of the buttocks [Figure 1–1]. The physician...takes into consideration whether the reduction should naturally be made straight downward, or towards the head, or towards the hip. This method of repositioning is harmless; indeed, it will do no harm even if one sits on the hump while extension is applied...nay there is nothing against putting one's foot on the hump and making gentle succession by bringing one's weight upon it [Figure 1–2].[1(p4)]

The description of lumbar extension for treatment of lumbar "kyphosis" is early testimony to some of the popular extension techniques for treatment of discogenic lesions in which a loss of lordosis is common. The idea of "repositioning" is definitely an early theme in the ancient documented literature on manual medicine.

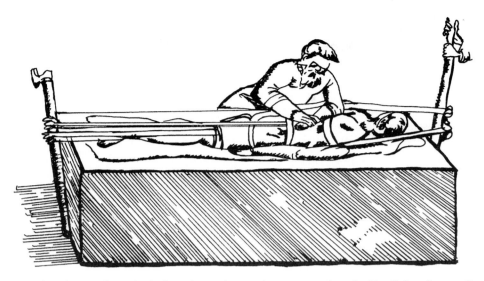

Figure 1–1 The Hippocratic method of traction and manual pressure as described by Galen. *Source:* Reprinted with permission from E.H. Schoitz, Manipulation treatment of the spinal column from the medical-historical standpoint, part I, *Journal of the Norwegian Medical Association* (1958;78:359–372), Copyright © 1958, Norske Laegeforening.

Figure 1–2 Method for "repositioning of an outward dislocation" of the spinal column. *Source:* Reprinted with permission from E.H. Schoitz, Manipulation treatment of the spinal column from the medical-historical standpoint, part I, *Journal of the Norwegian Medical Association* (1958;78:359–372), Copyright © 1958, Norske Laegeforening.

Claudius Galenus, or Galen, a Greek physician who lived in the years AD 129–199, contributed much written material on early manual medicine, including 18 commentaries on Hippocrates.[1] His primary contribution was documentation of early neurologic investigations. He recognized seven of the cranial nerves, differentiated between sensory and motor nerves, and was the first to treat paresthesias and extremity pain by treating the spine. Galen describes one such incident in which a patient developed paresthesias and loss of sensation in the third to fifth digits of the hand after falling from a wagon. Galen found that the problem was "localized in the first spinal nerve below the seventh cervical vertebra,"[1] and healed the patient by treating the neck. Much of the emphasis in Galen's work again focused on the "repositioning" of an outward dislocation of the spinal column.

While the advent of the Middle Ages brought a decline in medical advancement, an Arabic physician named Avicenna wrote a large work around the year AD 1000 summarizing the medical knowledge of the day. In the work, references are made to manual medicine, with descriptions and illustrations similar to the Hippocratic method. The Hippocratic method had survived, virtually unchanged in technique, well into the Middle Ages. It can be argued that many of the techniques (especially traction and extension principles) are still being utilized today.

Renaissance

The most well-known contributor to manual medicine in the Renaissance period was the French surgeon Ambroise Paré who lived in the 1500s.[1,4] Paré was also instrumental in the development of some of the early orthopedic surgical techniques. The positional theory was still strong as evidenced in a chapter entitled "Dislocated Spinal Vertebrae."

The exogenous causes of dislocation include falls, hard blows, and prolonged work in a greatly bent position, e.g. among vineyard workers…. If the vertebrae are dislocated and far apart, a good method is to lay the patient on a board, face down, fasten him to it with bands beneath his armpits, around his trunk and thighs, then pull from top and bottom as hard as possible, but without violence. If such tension cannot be tolerated, no treatment can be applied. Then you may place your hands on the outcurving part and press the projecting vertebrae.

Again, early evidence exists for traction and manipulation into extension, with the fundamental theory being repositioning of the vertebra as in the Hippocratic method.

Bone Setters

From the mid-1600s well into the nineteenth century, the "bone setters" of England flourished. Bone setters, considered "quacks" by traditional medical practitioners, had no formal training; their art was generally passed on from parents to children, generation after generation. Bone setters were known locally, had other primary occupations, and usually treated "con amore," that is, without pay.

Bone setters derived their name from their basic philosophy that small bones can move out of place, and healing takes place when the bones are restored to their original positions. One of the most well known bone setters was Sarah Mapp, a vagrant peasant woman, who was sought out by commoners and nobility alike (Figure 1–3). The fact that members of the nobility sought after bone setters infuriated the traditional medical community. For many years, the medical community hotly debated the subject of bone setting, with some physicians being shunned for speaking in favor of bone setters.

This controversy is exemplified by Wharton Hood, a medical doctor in the community, who learned the practice of bone setting from one of his patients whom he had treated for a systemic

Figure 1–3 The bone setter, Sarah Mapp (Crazy Sally). *Source:* Reprinted with permission from E.H. Schoitz, Manipulation of the spinal column from the medical-historical standpoint, part I, *Journal of the Norwegian Medical Association* (1958;78:359–372), Copyright © 1958, Norske Laegeforening.

illness. Realizing the effectiveness of such treatment in his own practice, Hood wrote boldly in the journals of the day in favor of bone setting.

> I obtained information, which surgeons do not learn, and which, if related to anatomical knowledge, is of the greatest possible value from the prophylactic and therapeutic viewpoints... It is entirely evident that quackery, among other things, is an expression of the extent to which the authorized physicians have failed to fulfill their patient's quite reasonable desires or demands. If the physician does not know how to fulfill or pursue these needs, it is his duty to study them, and in no respect can he fulfill his duty merely by criticizing quacks for his failures.[1(p5)]

Another physician of the day who defended manual medicine was English surgeon Sir James Paget, who was also a respected medical school professor (1814–1899). In a lecture to his students and later in an editorial to one of the medical journals he wrote:

> Few of you will enter into practice today without having a so-called bone setter as a competitor. There is little point in presenting a lecture on the injuries which these persons cause; it is more important to consider the fact that their treatment can do some good. . . .Learn then to imitate what is good and avoid what is bad in the practice of bone setters. Fas est ab hoste doceri! (It is advisable to learn from one's opponent.)[1(p6)]

Still another surgeon of the day wrote: "The success of certain bone setters is due—in addition to their skill—to the lack of practice and ignorance with which the practicing physician is equipped as concerns injuries to and diseases of the joints."

One of the best-known bone setters, Herbert Barker, who practiced from the late 1800s until 1927, vainly attempted to obtain credibility and good standing in the medical community by inviting physicians to observe his work and offering to perform demonstrations. His work was effective enough to attract members of the British royal family, actors, and politicians. Despite his successful treatments and his willingness to submit his work to the medical community's scrutiny, he was still shunned by the physicians of the day. Finally, frustrated by the arrogant attitudes of most physicians, Barker wrote: "Strong as the love of service to suffering is among many doctors as a whole, there exists some things much stronger and less worthy in prejudice and jealousy, which have from the beginning of time darkened the pages of surgical history, and smirched its record of noble endeavors."[5]

Eventually, the medical community could no longer argue with the success of bone setters, and in 1925 the *Lancet* editorially wrote: "The medical history of the future will have to record that our profession has greatly neglected this important subject.… The fact must be faced that the bone setters have been curing multitudes of cases by movement…and that by our faulty methods, we are largely responsible for their very existence."[6]

Osteopathic Medicine and Chiropractic

While controversy was raging over England's bone setters; a similar course of controversy was being charted in America during the 1800s and early 1900s. America's first bone setters were practicing by the mid-1800s in Rhode Island and Connecticut, and were criticized by skeptics just as in England.[4]

In the mid-1860s, Andrew Taylor Still, who had attended but never finished medical school, was helping his father cure native Indians and "simple folks" in the Midwest, when he lost three of his children to spinal meningitis. Disgusted with the traditional practice of medicine, he founded the practice of osteopathic medicine in 1874, probably influenced by the bone setters of his time. Taylor maintained that it was God who "asked him to fling in the breeze the banner of osteopathy." Being a very religious man, Still dedicated his first textbook to God: "Respectfully dedicated to the Grand Architect and Builder of the Universe."[7] His basic theory was that the human organism had the innate strength to combat disease, and as a vital machine of structure and function, would remain healthy as long as it remained structurally normal. If the structure was abnormal, the function would be adversely affected.[8] Still maintained that the causes of all diseases were "*dislocated bones*, abnormal, dislocated ligaments or contracted muscles, particularly in the spine, exercising a *mechanical pressure* on the blood vessels and nerves, a pressure that in part produces ischemia and necrosis, and in part an obstruction of the 'vital juices' through the nerves."[7] Thus, the rule of the artery and the rule of structure governing function became the cornerstones of osteopathic thought. Unfortunately, the treatment scheme included "cures" for all sorts of systemic diseases. Fortunately, osteopathic med-

icine continued to evolve into a more scientific and realistic philosophy. In 1956, the Register of Osteopaths in England compiled the Osteopathic *Blue Book*, which stated in part that "osteopathy is a system of therapeutics which lays chief emphasis upon the diagnosis and treatment of structural and mechanical derangements of the body."[8] By imposing these limitations, osteopathic physicians and osteopathic practice have become more accepted even though the theories are still debated. Three areas in osteopathic medicine that are currently applicable to myofascial manipulation are muscle energy techniques, positional release techniques, and strain/counterstrain techniques.[9-11]

In 1895, 21 years after Still had founded osteopathic medicine, David Daniel Palmer founded chiropractic. Some of the cure-all claims of osteopathic practice were being relinquished and were subsequently taken over by chiropractic. Palmer learned his technique through rediscovery of the ancient Hippocratic methods and from osteopathic medicine. He did, however, claim to be the founder of a new science.

> But I maintain to have been the first who repositioned dislocated vertebrae by using the spinous process and the transverse process as levers…and starting from these fundamental facts to have founded a science that is destined to revolutionize the theory and practice of the healing art.[7]

Dr. Charles Still, son of the founder of osteopathic medicine, maintained that Palmer had acquired his skills from a certain student at the Kirksville Osteopathic School and wrote that: "Chiropractic is the malignant tumor on the body of osteopathy."[7]

The original premise of chiropractic can be summed up as the "law of the nerve."

1. A vertebra can become *subluxated*.
2. A subluxation is apt to affect the structures that pass through the intervertebral foramen (nerves, blood vessels, and lymphatic vessels).

3. As a result thereof, a disruption of the function can occur at the corresponding segment in the spinal cord with its spinal and autonomic nerves, so that the conduction of nerve impulses becomes impaired.
4. As a result thereof, the innervations of certain parts of the organism change abnormally, so that they become functionally or organically sick, or they become disposed to disease.
5. An adjustment (*reposition*) of a subluxated vertebra causes the structures passing through the intervertebral foramen to be released, whereby the normal innervation of the organs is restored, so that they become functionally and organically rehabilitated.[7]

From ancient times to the end of the nineteenth century, manual medicine had been practiced with an apparent high degree of success. The emphasis during this time span was on repositioning a subluxation for the reduction of pain and restoration of health. With traditional medicine coming closer to embracing the value of manual medicine, the advent of the scientific age spurred new clinical investigations and research on the subject. Today, the subluxation philosophy has been partially replaced with the mobility philosophy in explaining the theories of manual medicine.

MODERN TIMES: THE TREND TOWARD MOBILITY AND DIAGNOSIS OF PATHOLOGY

In the early 1920s, physician participation in manual medicine became more common, especially in Great Britain, where the practice had been hotly debated for many years. One of the first physicians to publish extensively and authoritatively on the subject was Edgar Cyriax, father of the late James Cyriax. He is best remembered as one of the first to recognize discogenic pathology as a cause of back pain.[8,12]

In the late 1920s, the emergence of basic science and, especially arthrokinematics, became a significant factor in the study and philosophy of manual medicine. This influenced several others to explore further the theory of manual medicine. R.K. Ghormley was one of the first scientists to describe the facet joints as a possible cause of low-back pain.[13] He felt that arthritic changes in the facet joints narrowed the intervertebral foramen and were a possible cause of sciatic pain. Unfortunately, the condition he described was largely untreatable, and the hypothesis was later obscured by the idea of discogenic pathology as a cause of low-back pain and sciatica.[14] Basic science and arthrokinematics continued to influence and redefine manual medicine, however, and in the late 1940s and early 1950s, James Mennell published several volumes, including *Physical Treatment by Movement, Manipulation, and Massage*,[15] and *The Science and Art of Joint Manipulation*.[16] James Mennell was a strong advocate of intimate joint mechanics and the use of appropriate mobilization based on those same mechanics. He is believed to be the first to coin the term "accessory motion" to describe involuntary motions necessary in a joint for proper movement. He was also a strong proponent of the facet hypothesis in the evaluation and treatment of back pain, and recognized lack of mobility of the facet joints as being a causative factor in back pain. Mennell's early recognition of periarticular soft tissue dysfunction as a causative factor in back pain is significant in the development of the theoretical basis of soft tissue manipulation.

Also in the late 1940s and early 1950s, James Cyriax published the first edition of his now classic *Textbook of Orthopedic Medicine*.[17] The greatness of this publication lies in the differential diagnosis of musculoskeletal disorders and dysfunctions of the extremities. The work remains unsurpassed to this day. Cyriax's work is of special significance in the area of myofascial manipulation in the recognition, categorization, and differential diagnosis of the body's various soft tissues. The fact that pain could be caused by dysfunction of various or selective soft tissues, including, but not limited to, periarticular connective tissue, is a foundation of soft tissue manipulation today. Cyriax was also the first to introduce the concept of "end feel" in the diagnosis of soft tissue lesions. Cyriax summarizes his own philosophy as follows.

> In particular, I have tried to steer manipulation away from the lay notion of a panacea—the chief factor delaying its acceptance today. My only important discovery, on which the whole of this work rests, is the method of systematic examination of the moving parts by selective tension. By this means, precise diagnoses can be achieved in disorders of the radiotranslucent moving tissues.

The recognition of "radiotranslucent moving tissues" as the cause of pain is a cornerstone in the validation of treatment of soft tissue pathology, even though Cyriax deviated somewhat from his philosophy when evaluating and treating the spine. Oddly, his views on low-back pain remained strongly and narrowly in the realm of discogenic lesions, which is perplexing in light of the extremely systematic evaluation of the soft tissues advocated in extremity dysfunction.

Historically, the shift toward mobility and soft tissues in the etiology of back pain is quite evident by the mid-twentieth century. The trend continued with James Mennell's son, John, who was another advocate of the mobility philosophy. John Mennell operationally defined the different terms, which by this time had become confusing. In his book, *Joint Pain*, Mennell argued that the principal cause of pain arose from the synovial joints of the back, and not the disc.[18] He argued that there was no reason why the synovial joints of the spine should respond to trauma and/or therapeutic measures any differently from any other synovial joint of the body. Mennell outlined the etiological factors that give rise to joint pain:

1. Intrinsic joint trauma.
2. Immobilization that includes therapeutic immobilization, disuse, and aging.

3. The healing of a more serious pathological condition in the musculoskeletal system.

Mennell also advocated the following concepts in operationally defining manual therapy terminology.

1. There is a normal anatomical range of mechanical play movements in synovial joints. It is prerequisite to efficient pain free movement. This is joint play.
2. Loss of joint play results in a mechanical pathological condition manifested by impaired (or lost) function and pain. This is joint dysfunction.
3. Mechanical restoration of joint play by a second party is the logical treatment of joint dysfunction. This is joint manipulation.[19]

Thus, by moving joints in selective ways, the connective tissues surrounding the joint are appropriately stretched and normal movement is restored. The extensibility of the surrounding tissues is what ultimately allows for normal arthrokinematics in the joint.

Another person responsible for bringing arthrokinematics into the evaluation and treatment of joint pain was Norwegian physiotherapist Freddy Kaltenborn. Influenced by Cyriax, his classic text on extremity mobilization was the first that consistently and comprehensively used arthrokinematic principles to restore function to joints.[20] Kaltenborn was the first to advocate heavily the convex/concave rule for joint mobilization. He defined mobilization as "a component of manual therapy referring to any procedure that increases *mobility* of the soft tissues (soft tissue mobilization) and/or the joints (joint mobilization)."[20]

The implication made by Mennell, Cyriax, and others is that restoring the mobility of the joint restores normal function, and thereby reduces pain. A strong proponent of this idea was Stanley Paris, who wrote early on that the treatment of spinal pain involved treatment of the dysfunction, and not of the pain itself. "Dysfunction is the cause of pain. Pain follows dysfunction—pain cannot precede dysfunction. Pain does not warn of anything, it states 'something is wrong'."[3,21–23] By normalizing mobility and function in the spine, the pain would take care of itself. Paris further operationally defined the various accessory motions of joints in the following manner: (1) Component motions are those motions occurring in a joint during active motion, necessary for the motion to take place normally; and (2) joint play motions are those motions not under voluntary control, which occur only in response to outside forces.[21]

Paris developed a comprehensive evaluative system that included, in part, the evaluation of passive segmental mobility of the individual joints of the spine. He also classified manipulation into three distinct categories.

1. *Distraction:* when two articular surfaces are separated from one another. Distractions are used to unweight the joint surfaces, to relieve pressure on an intra-articular structure, to stretch a joint capsule, or to assist in the reduction of a dislocation.
2. *Nonthrust articulation:* when the joint is either oscillated within the limits of an accessory motion or taken to the end of its accessory range and then oscillated or stretched. Articulations are used mechanically to elongate the connective tissues, including adhesions, and neurophysiologically, to fire cutaneous, muscular, and joint receptor mechanisms.
3. *Thrust manipulation:* when a sudden high velocity, short amplitude motion is delivered at the pathological limit of an accessory motion. The purpose is either to alter positional relationships, snap an adhesion, or produce neurophysiological effects.[21]

Another recent proponent of the mobility theory is G.D. Maitland of Australia. His treatment system includes "graded oscillations" of grades I–IV (Figure 1–4). The oscillations are thought to work by increasing mobility as well as modulating pain through neurophysiological effects.

The mobility theory so began to dominate the thinking in manual therapy that, in the 1970s, the chiropractic profession redefined its philosophy to include movement abnormalities, while retaining its subluxation theory. Several recent studies have been performed using fluoroscopy to show changes in mobility of spinal facet joints after a thrust manipulation.[24] The studies are impressive and validate the effectiveness of manual therapy for increasing mobility.

Connective Tissue Research

The next logical step in the evolution of manual medicine was the emphasis on the histology and biomechanics of connective tissue. Since restoration of motion is manual therapy's primary goal, and since all the periarticular tissues affected during manual therapy are connective tissues (soft tissues), understanding the biomechanics of connective tissues became paramount. Substantial research was performed by Akeson, Amiel, Woo, and others to determine the biomechanical characteristics of normal and immobilized connective tissues. The findings of this research are discussed in detail in Chapters 3 and 4. Advances made in the understanding of connective tissue have helped explain manual therapy's effectiveness, especially myofascial manipulation. Others such as Kirkaldy-Willis

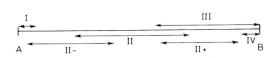

Figure 1–4 Grades of mobilization with **A** representing beginning movement, and **B** representing end-range movement. *Source:* Reprinted with permission from G.D. Maitland, *Peripheral Manipulation,* Woburn, Massachusetts, Butterworth-Heinemann, © 1981.

and Farfan have shed light on the degenerative pathologies in the spine, and have addressed the treatment of such conditions as well as some of the limitations of manual therapy.[25–26]

Future Considerations

Based on the current rate of change, manual therapy will continue to evolve exponentially into the twenty-first century. A significant addition to the realm of manual medicine is the idea of movement science. Although manual therapy can be effective in managing spinal problems, the incidence of recurrent spinal pain still borders on epidemic proportions. Integrating alternate somatic therapies such as Feldenkrais and Alexander and the theories of movement science with manual techniques makes sense in light of recurrent spinal pain, and takes the patient an extra step in prevention of recurrence. The idea of exercise for prevention of low-back pain is widely sanctioned, and conventional exercise can be considered movement science in rudimentary form. Manual technique can correct the dysfunction, and movement therapies help prevent future recurrence, creating a more complete form of treatment.

In addition, the idea that myofascial manipulation can produce not only mechanical and autonomic results, but also the modulation of central nervous system mechanisms, is in research infancy. The idea that myofascial manipulation can be a form of "sensory-motor education," helping to establish more efficient movement patterns will also strongly emerge to complement motor learning theories.[27]

REFERENCES

1. Schoitz EH. Manipulation treatment of the spinal column from the medical-historical standpoint. *J Norweg Med Assoc.* 1958:78:359–372.

2. Beard G, Wood E. *Massage: Principles and Technique.* Philadelphia: WB Saunders; 1964:3–4.

3. Loubert PV, Paris SV. *Foundations of Clinical Orthopedics.* St. Augustine, FL: Institute Press; 30–44.

4. Lomax E. *Manipulative Therapy: A Historical Perspective from Ancient Times to the Modern Era.* The Research Status of Spinal Manipulative Therapy. Bethesda, MD: National Institute of Neurological and Communicative Disorders and Stroke: 1975. Monograph 15.

5. Hood W. On the so-called bone setting, its nature and results. *Lancet.* 1871:336–338, 441–443, 499–501. (Taken from bibliography of note 1.)

6. Paget J. Cases that bone setters cure. *BMJ.* 1867. (Taken from bibliography of note 1.)

7. Schoitz EH. Manipulative treatment of the spine from a medical-historical point of view, II: Osteopathy and chiropractic. *J Norweg Med Assoc.* 1958:78:429–438.

8. Schoitz EH. Manipulative treatment of the column from the medical-historical point of view. III: The last 100 years. *J Norweg Med Assoc.* 1958:78:946–950.

9. Deig D. *Positional Release Techniques.* 1991. Course notes. Krannert Graduate School of Physical Therapy, University of Indianapolis, IN.

10. Jones L. Spontaneous release by positioning. *The D.O.* 1964:4:109–116.

11. Jones L. *Strain and Counterstrain.* Colorado Springs, CO: American Academy of Osteopathy; 1981.

12. Cyriax E. *Collected Papers on Mechano-Therapeutics.* London, England: Bale and Danielson; 1924. (Taken from bibliography of note 8.)

13. Ghormley RK. Low back pain with special reference to the articular facets. *JAMA.* l933:l0l:1773–l777.

14. Mixter WJ, Barr JS. Rupture of the intervertebral disc with involvement of the spinal canal. *New Engl Surg Soc.* 1934:2:210–215.

15. Mennell J B. *Physical Treatment by Movement, Manipulation and Massage.* Boston, MA: Little, Brown & Co; 1945.

16. Mennell JB. *The Science and Art of Joint Manipulation.* London, England: Churchill Ltd; 1949;52:I, II.

17. Cyriax J. *Textbook of Orthopedic Medicine.* Vol I, II. London, England: Baillière Tindall.

18. Mennell J McM. *Joint Pain.* Boston, MA: Little, Brown & Co; 1964.

19. Mennell J McM. *History of the Development of Medical Manipulative Concepts; Medical Terminology. The Research Status of Spinal Manipulative Therapy.* Bethesda, MD: National Institute of Neurological and Communicative Disorders and Stroke; 1975. Monograph 15.

20. Kaltenborn F. *Manual Therapy for the Extremity Joints*. Oslo, Norway: Olaf Norlis Bokhandel; 1976.

21. Paris SV. *The Spine—Etiology and Treatment of Dysfunction Including Joint Manipulation*. 1979. Course notes. Institute of Graduate Physical Therapy, St. Augustine, FL.

22. Paris SV. Mobilization of the spine. *Phys Ther*. 1979; 59(8):988–995.

23. Paris SV. Spinal manipulative therapy. *Clin Orthop*. 1983;179:5561.

24. Atlanta Craniomandibular Society/Life Chiropractic College Joint Seminar; August, 1987; Atlanta, GA.

25. Farfan HF. *Mechanical Disorders of the Low Back*. Philadelphia: Lea & Febiger; 1973.

26. Kirkaldy-Willis WH. *Managing Low Back Pain*. New York: Churchill Livingstone; 1988.

27. Juhan D. *Job's Body: A Handbook for Bodywork*. Barrytown, NY: Station Hill Press; 1987.

CHAPTER 2

Modern Theories and Systems of Myofascial Manipulation

Robert I. Cantu and Alan J. Grodin

This chapter provides an overview of some of the alternate somatic therapies considered myofascial in nature. Its purpose is neither to give the reader a comprehensive background of each individual system, nor to include every system currently being practiced—such an undertaking is a book in itself. The systems reviewed represent those that have influenced the authors the most over the years, and have contributed to the development of the authors' personal treatment philosophies. The manual therapist interested in myofascial manipulation should also have a basic working knowledge of the fundamental philosophies behind various systems and theories in order to become a more educated consumer in the continuing education market, and to understand the orientation of the respective practitioners.

Modern theories and systems are arranged in three categories: autonomic or reflexive approaches, mechanical approaches, and movement approaches. Autonomic approaches are those that exert their therapeutic effect on the autonomic nervous system. Mechanical approaches are those that actually attempt mechanical changes in the myofascia by direct application of force, and movement approaches are those that attempt to change aberrant movement patterns and establish more optimal ones. Ideally, the manual therapist should have a basic working knowledge of theories or systems in all three areas, along with some application technique from each approach.

AUTONOMIC APPROACHES

The autonomic or reflexive approaches attempt to exert their effect through the skin and superficial connective tissues.[1,2] MacKenzie defined the autonomic or reflexive component as "that vital process which is concerned in the reception of a stimulus by one organ or tissue and its conduction to another organ, which on receiving a stimulus produces the effect."[3] Soft tissue mobilization performed for autonomic effect stimulates sensory receptors in the skin and superficial fascia. These stimuli pass through afferent pathways to the spinal cord and may be channeled through autonomic pathways, producing effects in areas corresponding to dermatomal zones being mobilized.[4]

The idea of affecting various body areas by stimulating the skin and superficial connective tissue has been used in areas apart from soft tissue mobilization. For example, part of the theory of transcutaneous electrical nerve stimulation (TENS) is direct stimulation of large myelinated nerve fibers that override noxious stimuli traveling to higher centers of the central nervous system. So, TENS has application not only for pain control, but also for control of post-surgical nausea or menstrual cramping.

Affecting the autonomic system is an important steppingstone to more aggressive, mechanical work, especially in acute patients. In subacute patients, autonomic techniques are most often used at the beginning and at the end of treatment to provide entry and exit from mechanical technique. The effects of autonomic technique should not be overemphasized, however. Some practitioners have used this autonomic phenomenon to justify treatment of disorders unrelated to the neuromusculoskeletal system. Although the autonomic effect cannot be denied, judgment should be exercised by the clinician in evaluating the extent of autonomic treatment.

Connective Tissue Massage (Bindegwebbsmassage)

Connective tissue massage (CTM) was developed in the 1920s by German physiotherapist Elizabeth Dicke[1] and later expanded by Maria Ebner.[2] The system was conceptualized and put into practice in rudimentary form in the late 1920s when Dicke was suffering from a prolonged illness caused by an "impairment of the circulation" in her right leg, later diagnosed as endarteritis obliterans. The leg was cold and discolored with significant, but diffuse, numbness, and loss of distal pulses. The attending physicians prescribed a prolonged period of bed rest. If the bed rest was unsuccessful in diminishing the symptoms, amputation would have been considered as a last resort. Dicke was in bed for a 5-month period, during which time she understandably developed significant low-back pain. As she began to palpate her own back, she found exquisite tenderness, hypersensitivity, and palpatory changes in the area of the iliac crest and sacrum. She stated that she felt "a thickened infiltrated area of tissue, and opposite it, an increased tension of the epidermis and dermis."[1] She found relief by gently and superficially stroking the area with her fingertips. Over time the low-back pain diminished, but more important, notable changes occurred in the lower extremity. She initially felt itching, followed by warm flushes and increased sensation. She then began exploring the extremity itself and found other hypersensitive areas, especially along the border of the greater trochanter and the iliotibial tract. She very gently and superficially stroked these areas, and improvement continued. Within 3 months her symptoms had subsided, and shortly thereafter, she was able to resume her full duties as a physiotherapist.

Out of her experiences, she gradually constructed a systematic treatment method. From this pursuit, she also extrapolated a treatment of organic pain, which is beyond the realm of this book. The effects Dicke outlined that are pertinent to modern manual therapy are as follows.

1. CTM can directly influence connective tissue that is locally altered by illness, i.e., scars, local blood supply, and other disturbances.
2. CTM can set general circulation in order. Subcutaneous connective tissue is extremely vascularized and can absorb varied quantities of blood as a result of constriction or dilation.
3. CTM can also release nerve impulses along quite specific paths by means of reflexes that are locked into the central nervous system. It can create reactions in distant organs. Dicke refers to certain aspects of this phenomenon as the "cutivisceral reflex."[1] Dicke uses the example of the application of a mother's warm hand to alleviate a child's stomachache. Obviously, the intestine would not be affected from the surface of the skin and the reaction must be "a reflex which affects the intestines from the skin" (Figure 2–1).[1] The skin and subcutaneous tissues, which are highly innervated and vascularized, are the primary tissues for the reception of outside tactile stimuli.

The CTM system is very systematic and protocol-oriented if performed as Dicke taught. Each stroke, for example, is performed three times, with the right side always first. Most

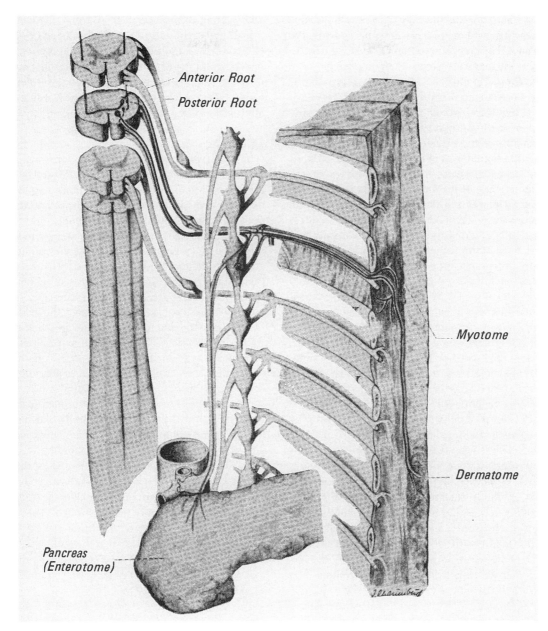

Figure 2–1 An example of the cutivisceral reflex as described by Dicke. *Source:* Reprinted from *Segmentale Innervation* by K. Hansen and H. Schliack with permission of Georg Thieme Verlag, © 1962.

strokes are performed with the middle finger of the hand, with the other hand always in light contact with the patient. Lubrication is never used, and the low back and sacral areas are always treated first. Treatment is never administered without first treating the basic section of the low back, sacrum, and coccyx, with a "build up" to the affected area. What must be remembered about CTM and about all other "systems" is that they are merely systems. Astute clinicians can and should modify these systems while assisting their patients to recovery.

CTM exerts its effect using the skin and subcutaneous connective tissue. This makes CTM primarily a superficial form of myofascial manipulation (in terms of depth of penetration); one that provides much-needed techniques on the "lighter" end of the manual technique spectrum. Manual therapists often move too quickly into moderate or deep-level technique, instead of gradually entering the myofascial system.

CTM offers other therapeutic advantages when integrated properly into the overall treatment scheme. In a patient who is autonomically hypersensitive, CTM provides the type of technique that can quiet the system. Such an acute patient can be described as having an RSD- (reflex sympathetic dystrophy) type back. Often seen in the hands and feet, RSD is a hyperactivity of the sympathetic nervous system that creates chronic intense pain and hypersensitivity, cold clammy feeling, cold sweat in the area, nausea with attempted palpation, and eventually trophic changes including shiny skin and bone and hair loss. A patient with an RSD-type back may display some of these symptoms, although without most of the accompanying signs. The patient may exhibit hypersensitivity to palpation, a cold clammy feel to the back with attempted palpation or treatment, and a nausea response. The superficiality of CTM makes it a prime choice of technique, since it primarily affects the autonomic nervous system. CTM technique also allows the patient to grow accustomed to the clinician's hands in a very nonthreatening manner, further promoting relaxation and rapport.

In cases where deeper myofascial restrictions exist, CTM technique provides a good entry into the deeper tissues. If the clinician penetrates the layers of myofascia too rapidly, reflex guarding of the deep myofascia may result, rendering treatment more difficult. Moving from superficial to deep treatment facilitates the accessing and treating of the deeper myofascial elements. The clinician thus allows the body to open itself to treatment, which becomes less forceful with less potential for tissue microtrauma and exacerbation of pain. Deep technique does not need to be aggressive, if the deep tissue is accessed appropriately.

Hoffa Massage

Albert Hoffa's text, published in 1900 and later revised by Max Bohm in 1913, represents more classical massage techniques such as effleurage, pétrissage, tapotement, and vibration.[4-6] Most therapists learn these as standard massage techniques in entry-level programs, but they should still be recognized and discussed because of their importance in the overall treatment scheme. Some may disregard this type of massage, regarding it as too basic to include in the realm of advanced manual therapy; but leaving behind traditional myofascial manipulation techniques can handicap even the most advanced manual therapists. A technique is not necessarily more effective just because it is more complex.

Some may consider these techniques to be more mechanical in nature, but the strokes can be lighter and designed to be relaxing, which categorizes them as reflexive or autonomic. Many myofascial manipulation systems are neither exclusively reflexive nor exclusively mechanical, but may lean toward one more than the other. Hoffa massage certainly inclines toward the reflexive. Hoffa states that "the force should be gentle and 'light-handed' so that the patient feels as little pain as possible."[5] Hoffa advocates that massage should never last more than 15 minutes, even for the whole body.

As with connective tissue massage, Hoffa's concept emphasizes using autonomic or reflexive technique as an entryway for other, more mechanical technique. With Hoffa massage or CTM, the patient is prepared, and more specifically, the myofascia is prepared for techniques designed to promote histological changes in the myofascial tissues. The changes can be made without forceful maneuvers that can create microtrauma or exacerbate painful conditions. Some of Hoffa's basic massage strokes are described as follows.

Light and deep effleurage. The hand is applied as closely as possible to the part. It glides on it, distally to proximally. . . . With the broad part of the hand, use the ball of the thumb and little fingers to stroke out the muscle-masses, and at the same time, slide along at the edge of the muscle with finger tips to take care of all larger vessels: stroke upward.

One-hand pétrissage. Place the hand around the part so that the muscle-masses are caught between the fingers and thumb as in a pair of tongs. By lifting the muscle mass from the bone "squeeze it out," progressing centripetally.

Two-hand pétrissage. Apply both hands obliquely to the direction of the muscle fibers. The thumbs are opposed to the rest of the fingers. This manipulation starts peripherally and proceeds centripetally, following the direction of the muscle fibers. The hand that goes first tries to pick the muscle from the bone, moving back and forth in a zigzag path. The hand that follows proceeds likewise, "gripping back and forth." . . . On flat surfaces where this petrissage is not possible, . . . stroke using a flat hand, instead of picking up the muscle.

Tapotement. Both hands are held vertically above the part to be treated in a position that is midway between pronation and supination. Bringing them into supination, the abducted fingers are hit against the body with not too much force and with great speed and elasticity. Fingers and wrists remain as stiff as possible but the shoulder joint comes into play all the more actively.[4]

Hoffa was one of the first clinicians to describe massage in an actual textbook. The fundamental strokes of traditional massage are still performed widely today, although many variations have been introduced. Hoffa's massage is considered basic by modern standards, but advanced manual therapists continue to use his techniques in their treatment schemes.

MECHANICAL APPROACHES

Mechanical approaches differ from autonomic approaches in that they seek to make mechanical, or histological, changes in the myofascial structures. The stretching of a hamstring, the elongation of a superficial fascial plane, or superficial tissue rolling to mobilize adhesions are all mechanical techniques. As previously stated, mechanical techniques should generally be performed after some form of autonomic technique. Even if the patient is not suffering acute pain, a few minutes of autonomic technique facilitate the application of mechanical technique. The application of mechanical technique is not necessarily aggressive; it is a matter of properly going through the "layers" until the deeper tissues are accessed. That is not to say that aggressive, forceful mechanical technique is an inferior form of treatment; at times, forceful technique is necessary to free up longstanding restrictions. The gentle, however, should always be attempted first.

Remember that the systems described as follows are just that: systems—they can be very protocol-oriented, and very ordered. Principles may be borrowed from any system, however, and may be effective if used at the proper time and in the proper sequence.

Rolfing® (Structural Integration)*

Structural integration, a system created by Ida Rolf, is used to correct inefficient posture

---

*Rolfing® is a registered service mark of the Rolf Institute of Structural Integration.

or to integrate structure. The technique involves manual soft tissue manipulation with the goal of balancing the body in the gravitational field (Figure 2–2). Rolfing is a standardized, non-symptomatic approach to soft tissue manipulation, administered independent of specific pathologies.

The technique involves 10 one-hour sessions, each emphasizing a particular aspect of posture, with all the work performed in the myofascial tissues. Two or three advanced sessions can be performed, as well as subsequent occasional "tune-up" sessions. The treatment principle says that "if tissue is restrained, and balanced movement demanded at a nearby joint, tissue and joint will relocate in a more appropriate equilibrium" (Figure 2–3).[7] The ten sessions include

1. Respiration
2. Balance under the body (feet/legs)
3. Lateral line-front to back (sagittal plane balance)
4. Base of body/midline (balance left to right)

5. Rectus abdominis/psoas-for pelvic balance
6. Sacrum-weight transfer from head to feet
7. Relationship of head to rest of body-primarily occiput/atlas (OA) relationship, then to rest of body
8, 9. Upper and lower half of body relationship
10. Balance throughout system[8]

Rolfing also strives to integrate the structural with the psychological:

> The technique of Structural Integration deals primarily with the physical man; in practice, considerations of the physical are inseparable from considerations of the psychological. . . . Emotional response is behavior, is function. All behavior is expressed through the musculoskeletal system. . . . A man's emotional state may be seen as the projection of his structural imbalances. The easiest, quickest and most economical method of changing

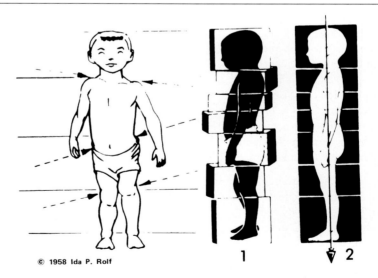

© 1958 Ida P. Rolf

1   **2**

Figure 2–2 The concept of balancing posture in a gravitational field, with the body consisting of various blocks. *Source:* Reprinted from *Rolfing: The Integration of Human Structures* (p 33) by I. Rolf with permission of the Rolf Institute of Structural Integration. © 1977.

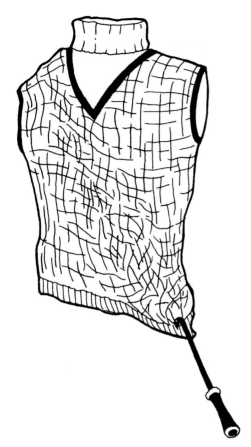

Figure 2–3 The fascial sweater concept showing that a fascial restriction in one area will strain areas away from the restriction and cause abnormal movement patterns. *Source:* Reprinted from *Rolfing: The Integration of Human Structures* (p 33) by I. Rolf with permission of the Rolf Institute of Structural Integration. © 1977.

the coarse matter of the physical body is by direct intervention in the body. Change in the coarser medium alters the less palpable emotional person and his projections.[7]

Rolfing suggests that a person's psychological components are manifested in structure, and that changing the structure can change the psychological component.

Trager®*

Tragering is a mechanical soft tissue and neurophysiological reeducation approach developed gradually over the last 50 years by Milton Trager, MD. The approach has no rigid procedures or protocols like some other systems. It uses the nervous system to make changes, rather than making mechanical changes in the connective tissues themselves. The Trager practitioner "uses the hands to communicate a quality of feeling to the nervous system, and this feeling then elicits tissue response within the client."[9] Trager began developing his system in his late teens, while training as a boxer. He subsequently left boxing to protect his hands and to pursue the development of his system. Eight years later, Trager undertook formal medical training, earning his medical doctorate at the University Autonoma de Guadalajara in Mexico. He opened his private practice in 1959 in Waikiki and, in the early 1970s, began teaching his system on an individual basis in California. The Trager Institute was formed and there are currently 600 Trager practitioners throughout the world.

Tragering is directed toward the unconscious mind of the patient: "for every physical non-yielding condition there is a psychic counterpart in the unconscious mind, and exactly to the degree of the physical manifestation."[10] The system uses gentle passive motions that emphasize mobilization techniques, concentrating on traction and rotation, and a system of active movements termed Mentastics®. The intensity of the movements is in the moderate or midrange, with integration of cervical and lumbar traction. The oscillations and rocking techniques serve as relaxation techniques that encourage the patient gradually to relinquish control. Finally, the active movement part of the treatment serves as a neuromuscular reeducation technique similar in principle to Feldenkrais' work. The idea is to alter the patient's neurophysiological set and give the patient the tools to maintain the

---

*Trager® is a registered service mark of the Trager Institute.

changes.[11] The therapist is not attempting to make mechanical changes in the soft tissues, but is trying to alter the neuromuscular set to establish more normal movement patterns.

MOVEMENT APPROACHES

The movement approaches differ from the others in that the patient actively participates in therapy. Both autonomic and mechanical approaches rely on the clinician to impart the changes and movement. In the movement approaches, the clinician guides the patient through a series of movements to change aberrant patterns and retrain into more efficient movements and postures.

Alexander

F. Matthias Alexander was a Shakespearian orator at the turn of the twentieth century. He developed a consistent problem in projecting his voice. He began studying the relationship of head and neck posture in relation to voice projection, and from that developed a system of movement that can teach the entire body to become more efficient, regardless of the activity. The technique objectives are improvements in both posture and body mechanics. Many vocalists, musicians, and other performing artists use the Alexander technique to improve efficiency. Since Alexander's recurrent laryngitis persisted despite prolonged periods of rest, he set up a system of mirrors through which he could observe himself speaking in his professional oratorical voice. He observed a tendency to pull his head back, depress his larynx, and inhale through his mouth. After repeated practice sessions, he was able to hold his head and neck in a more efficient posture, and with time, his voice projection improved and his laryngitis subsided. As time passed, Alexander noticed that the "dysfunctional" head tilting was not an isolated movement, but was coordinated with other dysfunctional patterns throughout his body such as lifting the chest, arching the back, and tensing the legs and feet.[12] Alexander theorized

that in each human being, there existed an integrating mechanism that produced more optimal coordination and functioning. Alexander wrote:

> I discovered that a certain use of the head in relation to the neck, and of the head and neck in relation to the torso and other parts of the organism . . . constituted a primary control of the mechanisms *as a whole* . . . and that when I interfered with the employment of the primary control of my manner, this was always associated with a lowering of the standard of my general functioning.[13]

Position and motion of the head and neck are, therefore, the cornerstones of the Alexander technique. The student of Alexander learns to activate this primary locus of control in the head and neck, and keep it functioning during activities of daily living.

The instructor's approach is usually to give the student palpatory as well as verbal feedback as he or she learns new positions and movement patterns. As the student masters new patterns, less palpatory and verbal feedback is given, until the student can independently achieve proper control patterns. Alexander was very experiential and deliberate in his approach, reasoning that, like music teachers who suggest that their students practice slowly, patterns are best learned slowly and with positive reinforcement.

In learning or teaching the technique, Alexander goes through three stages: (1) awareness of the habit; (2) inhibition of the habit; and (3) conscious control of the habit. These three stages are what Alexander termed "conscious learning," where the participant deliberately and actively tries to change old habits while incorporating new ones.

Awareness of the habit carries great importance in the Alexander technique: "You are not here to do exercises, or to learn to do something right, but to be able to meet a stimulus that always puts you wrong and learn to deal with it."[12] For Alexander, his public speaking triggered the dysfunctional patterns. He found he had difficulty even recognizing the patterns that were so detrimental to his voice projection. He hypothesized that the brain no longer identified the aberrant patterns of movement as dysfunctional, but as normal. Simply looking in the mirror to correct an aberrant postural or movement dysfunction was insufficient to change the pattern. Developing an awareness of the pattern was the first step.

Once the dysfunctional pattern was recognized, inhibition of the movement was necessary, but again, being aware of the pattern was not enough to change it, since the habit was too well established. He began to speak while consciously trying to "turn off" the dysfunctional pattern. He then used conscious control to "inhibit" the dysfunctional pattern and integrate the new one.

Some of these principles are integrated into sequencing of overall treatment. If a patient exhibits poor posture resulting from myofascial restrictions and movement imbalances, mechanical approaches are used to free up the restrictions, allowing the patient to assume optimal posture without undue effort. If new posture is emphasized too early in the treatment sequence, the patient often may not have the body awareness or the ability to assume it. The new posture, then, can increase the patient's original pain, and establish a negative reinforcement loop. If the clinician addresses mechanical restrictions and emphasizes body awareness, the patient becomes aware of the problem, is able to inhibit the old pattern, and consciously work toward establishing the new pattern, with more efficient effort.

Alexander's concepts have been used and expanded by Mariano Rocobado, Steve Kraus, and others in working with head and neck posture in relation to mandibular position. As is widely known, head and neck posture and movement affect mandibular position and function; the Alexander technique aptly applies to the evaluation and treatment of temporomandibular joint (TMJ) disorders. Whether used for treatment of TMJ, neck, or other spinal dysfunctions, the Alexander technique merges logically with the autonomic and mechanical approaches in helping myofascially dysfunctional patients achieve desired changes.

Feldenkrais

The Feldenkrais movement approach seeks to retrain the body away from aberrant movement patterns into more efficient ones. Moshe Feldenkrais was a versatile Israeli engineer and physicist who was also athletically active. Feldenkrais participated in soccer and judo, but a persistent knee injury resulting from soccer play led his engineering mind to explore human movement. His movement approach is based on the idea that movement abnormalities occur in response to past trauma, rendering one more susceptible to reinjury. His approach is designed to help the body reprogram the brain to integrate the whole mind-body entity.

Feldenkrais has two basic approaches, which he separates only for convenience. The first is an experiential approach that he terms "Awareness Through Movement,"[14] in which the patient receives a series of verbal commands designed to weaken old movement patterns and to establish new ones. The second is a hands-on approach that he terms "Functional Integration."[15] Feldenkrais disliked separating the two, especially if:

> ...the distinction is made that one is for "sick" or "brain damaged" people, and the other is for "normal, healthy" people. Which of us, after all, is not brain damaged in the sense that we allow many areas of our brains to atrophy through misuse or nonuse? We can have terrible posture and movement patterns and habits which are distorting and damaging to our bodies and brains—and still be classified as "normal." Who are we, then to call other people brain damaged simply because their particular deficiency produces visible effects that we label "disease?"[16]

The idea that all persons exhibit some abnormal movement either from previous trauma or old habit patterns is a cornerstone of the Feldenkrais method. As with Alexander technique, gentle sequences of movement allow for slow, deliberate changing of abnormal, inefficient movement patterns into normal efficient movements.

CONCLUSION

Examples of the three types of approaches (autonomic, mechanical, and movement) described here merge well with the authors' philosophy and scheme of treatment. As will be seen in later chapters, the sequencing of treatment includes beginning superficially with a manual approach, and working gradually into deeper tissues. Once the deeper tissues are accessed and affected, elongation of the structures becomes facilitated. When optimal length and mobility are established, neuromuscular reeducation is emphasized to prevent recurrence, as well as postural integration. The progression from a light manual approach (autonomic) to a deep manual approach (mechanical), and then to an emphasis in movement and posture (movement approach) is the key to complete treatment.

REFERENCES

1. Dicke E, Schliaek H, Wolff A. *A Manual of Reflexive Therapy of the Connective Tissue.* Scarsdale, NY: Sidney S Simon Publishers; 1978.

2. Ebner M. *Connective Tissue Manipulations.* Malabar, FL: Robert E Kreiger Publishing Co, Inc; 1985.

3. MacKenzie J. *Angina Pectoris.* London: Henry Frowde and Hodder and Stroughton; 1923:47.

4. Tappan EM. *Healing Massage Technique: A Study of Eastern and Western Methods.* Reston, VA: Reston Publishing Co; 1978:17–22.

5. Hoffa AJ. *Technik der Massage.* 14th ed. Stuttgart, Germany: Ferdinand Enke; 1900.

6. Bohm M. *Massage: Its Principles and Technique.* Philadelphia: WB Saunders; 1913.

7. Rolf IP. *Rolfing: The Integration of Human Structures.* Rochester, VT: Healing Arts Press; 1977.

8. Gordon P. *Myofascial Reorganization.* Course notes. 1988. The Gordon Group, Brookline, MA.

9. Juhan D. The Trager approach-psychophysical integration and mentastics. *The Trager Journal.* Fall 1987:1.

10. Trager M. Trager psychophysical integration and mentastics. *The Trager Journal.* Fall 1982:5.

11. Witt P. Trager psychophysical integration: an additional tool in the treatment of chronic spinal pain and dysfunction. *Whirlpool.* Summer 1986.

12. Rosenthal E. The Alexander technique—what it is and how it works. *Medical Problems of Performing Artists.* June 1987:53–57.

13. Alexander FM. *The Universal Constant in Living.* New York: Dutton; 1941:10.

14. Feldenkrais M. *Awareness through Movement.* New York: Harper & Row; 1972.

15. Rywerant Y. *The Feldenkrais Method: Teaching by Handling.* San Francisco: Harper & Row; 1983.

16. Rosenfeld A. Teaching the body how to program the brain is Moshe's 'miracle'. *Smithsonian.* January 1981.

PART II

Scientific Basis for Myofascial Manipulation

CHAPTER 3

Histology and Biomechanics of Myofascia

Robert I. Cantu and Deborah Cobb

The foundations of orthopedic physical therapy are based upon the understanding of the anatomy and biomechanics of the soft tissues. A manual physical therapist must have in-depth knowledge of the microscopic and macroscopic structure of the myofascial tissue—connective tissue, muscle, and junctional zones. This is essential because the myofascial/connective tissues are those primarily affected by manual therapy treatments. Thorough knowledge of myofascial tissue histology and biomechanics will aid the physical therapist in comprehending and assessing the implications of trauma, immobilization, and remobilization of myofascial tissues.

HISTOLOGY AND BIOMECHANICS OF CONNECTIVE TISSUE

Connective tissue comprises 16 percent of a person's total body weight and stores 23 percent of the body's total water content.[1] Connective tissue forms the base of the skin, the muscle sheaths, nerve sheaths, tendons, ligaments, joint capsules, periosteum, aponeuroses, blood vessel walls, and the bed and framework of the internal organs.[1,2] Also, from a histological standpoint, bone adipose and cartilage are considered connective tissues. The most important roles of connective tissue are (1) structural, due to the mechanical properties; and (2) defensive/reconstructive, in that they aid against invading microorganisms and contribute to repair after injury.[3] The importance of these roles to the manual therapist will be discussed later.

Most of the structures affected by manipulation and mobilization are connective tissues. When mobilizing a facet joint, for example, the tissue affected by the mobilization technique is the joint capsule, the surrounding periarticular connective tissue, nearby ligaments, and fascia. The joint is simply a space built for motion, but it is the surrounding connective tissues that are affected by the mobilization.

An appropriate understanding of normal histology and biomechanics of the connective tissues can be found in a review of the scientific literature. Although much of the benchmark research is from earlier in the century, it remains accurate and consistent with the more current research. This information will begin to lay the groundwork for an understanding of how trauma, immobilization, and remobilization will affect the connective tissues.

Histology

The four basic types of tissue found in the human body are muscle, nerve, epithelium, and connective tissue.[2] Connective tissue is subclassified into connective tissue proper, cartilage, and bone. Connective tissue proper is further subclassified by orientation and density of fiber types.[4] The three basic connective tissue types are dense regular, dense irregular, and loose irregular (Figure 3–1).[4] These tissue types are described in detail later in this chapter.

The Cells of Connective Tissue

Connective tissue is comprised of cells and extracellular matrix (fibers and ground substance; Table 3–1). These cells can be divided up into a fixed cell population of fibroblasts, adipocytes, persistent mesenchymal stem cells,

and mobile wandering cells consisting of macrophages, lymphocytes, plasma cells, eosinophilic leukocytes, and mast cells.[5] Fibroblasts are found in all connective tissues, whereas the other cells are found primarily in pathological states.

Fibroblasts. Fibroblasts, considered the true connective tissue cells, are found in the highest cell numbers. These cells are the primary secretory cells in connective tissue and are respon-

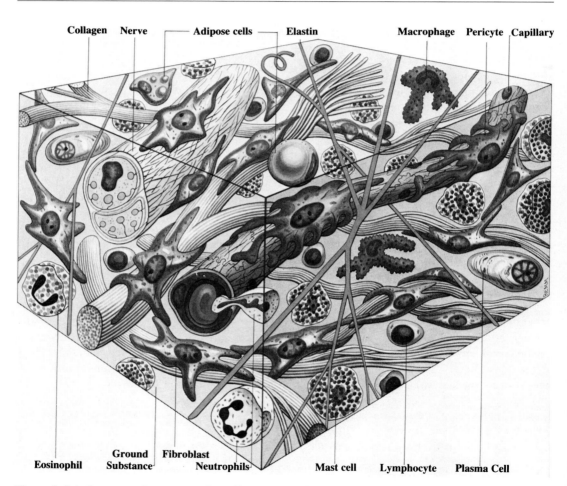

Figure 3–1 A diagrammatic representation of loose connective tissue, showing fibers, cells, ground substance, nerve, and blood vessels. *Source:* Reprinted from *Gray's Anatomy*, ed 35 (p 32) by P. Williams and R. Warwick with permission of W.B. Saunders, © 1973.

Table 3–1 Histological Makeup of Connective Tissue

I. Cells
  A. Fibroblasts: synthesize collagen, elastin, reticulin, and ground substance.
  B. Fibrocytes; mature version of fibroblast, found in stable mature connective tissue.
  C. Macrophages and histiocytes: "big eaters" found in traumatic, inflammatory, or infectious conditions. Clean and debride area of waste and foreign products.
  D. Mast cells: secrete histamine (vasodilator) and heparin (anticoagulant).
  E. Plasma cells: produce antibodies; present only in infectious conditions.
II. Extracellular Matrix
  A. Fibers
    1. Collagen: very tensile
      a. type I: connective tissue proper (loose and dense)
      b. type II: hyaline cartilage
      c. type III: fetal dermis, lining of arteries
      d. type IV: basement membranes
    2. Elastin: more elastic, found in lining of arteries. Also ligamentum flavum and ligamentum nuchae.
    3. Reticulin: delicate meshwork for support of internal organs and glands.
  B. Ground substance: viscous gel with high water concentration. Provides medium in which collagen and cells lie.
    1. Purpose
      a. diffusion of nutrients and waste products
      b. mechanical barrier against bacteria
      c. maintains critical interfiber distance, preventing microadhesions
      d. provides lubrication between collagen fibers
      e. more abundant in early life; decreases with age
    2. Components
      a. glycosaminoglycans (GaGs): lubricating effect, maintenance of critical interfiber distance, etc
      b. proteoglycans: primarily bind water

sible for the synthesis of all components of connective tissue, including collagen, elastin, and ground substance. Fibroblasts are adherent to the fibers, which they lay down. In highly cellular tissues, fibroblasts may mix with collagen fibers to become reticular cells.[3] In mature stable connective tissue, the fibroblast is converted into the fibrocyte, which is the nonsecretory version of the fibroblast. Fibroblasts and fibroblastic activity are influenced by various factors, including prevalent mechanical stresses, steroid hormone, and dietary content. Fibroblasts are nonphagocytic.

Macrophages. Other types of cells, not exclusive to connective tissue, are found primarily in traumatized or infectious states. Macrophages (which means "big eater") are responsible for phagocytosing waste products, damaged tissue, and foreign matter. In traumatized states, macrophages primarily phagocytose damaged cells and damaged macromolecular connective tissue fibers, debriding the area in preparation for repair. In infectious or inflammatory states, macrophages are capable of phagocytosing bacteria or other invading microorganisms.[3] Macrophages may be the signal for vascular regeneration to begin.

Mast cells. Mast cells were given their name because they appeared "stuffed with granules" (mast is German for well-fed). They are mobile and are important defensive cells, which are formed primarily in loose connective tissue. Mast cells are responsible for constantly secreting small amounts of the anticoagulant heparin. Heparin is constantly secreted in small amounts in the blood stream by the mast cells. The significance of this is still not known.[5] The disruption of mast cells also results in the release of histamine. Within the mast cell granules, histamine is bound to heparin. Histamine causes vasodilation in neighboring noninjured vessels, resulting in increased permeability. The release of histamine is linked to inflammatory reactions, allergies, and hypersensitivities.[1–5]

Mast cells can be hypersensitized by certain antigens introduced into the body, facilitating cell production of histamine.[2] This could be one possibility why individuals with numerous allergies and with diffuse myofascial pain can have an increased histamine response to soft tissue manipulation. This concept is discussed again later in the chapter on myofascial pain syndromes. Plasma cells are somewhat related to mast cells in that they are primarily present in infectious states. They are related to the immune system and are responsible for synthesizing antibodies.

Other connective tissue cells. With the exception of the fibroblast and fibrocyte, all other cells found in connective tissue are also related to the reticuloendothelial system. This widely scattered system consists of phagocytic and immunologic cells and associated organs and tissues related to first-line defense of the body against invading microorganisms and foreign particles.[3] Aside from connective tissue, the cells of the reticuloendothelial system are found in the blood, and the reticular tissue of the spleen, liver, and the meninges. The body's connective tissue framework is an integral part of the reticuloendothelial system because of the mechanical barrier that connective tissue provides against invading microorganisms.

The Extracellular Matrix

The extracellular matrix of connective tissue comprises all other components of connective tissue except cells (Table 3–1; Figure 3–2). The matrix is primarily composed of *fibers* and *ground substance*. The fiber types consist of collagen, elastin, and reticulin. Collagen, the most

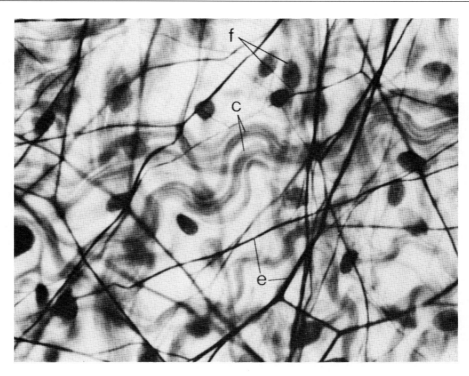

Figure 3–2 Photomicrograph of loose connective tissue. The connective tissue fibers lie in a bed of ground substance. *Source:* Reprinted from *Histology* (p 212) by A.W. Ham and D.H. Cormack with permission of J.B. Lippincott Co., © 1979.

commonly found fiber, is very tensile, whereas elastin and reticulin are more elastic. It is primarily the properties of the inert extracellular matrix that account for the functional characteristics of the different types of connective tissue. Connective tissue fibers with their tensile strength and elasticity are the basis for the mechanical support. Ground substance, with its water binding capacity, is the basis for lubrication and diffusion of nutrients in connective tissues.[5]

Collagen is divided into four major types: Type I collagen is found primarily in ordinary loose and dense connective tissue; Type II collagen is found primarily in hyaline cartilage; Type III collagen is found lining the fetal dermis; and Type IV collagen is found in basement membranes. Manual therapy techniques are most likely affecting Type I collagen. The characteristics of each type are discussed later.

Elastin fibers are less tensile than collagen and have more elastic characteristics. The lining of arteries contains a high percentage of elastin. The ligamentum nuchae of the spine is a ligament that contains a high percentage of elastin.[6,7] Reticulin is the least tensile of the connective tissue fibers; it is found primarily in the delicate meshwork supporting the body's internal organs and glands.

Another important component of connective tissue is *ground substance*. This is the viscous, hydrophilic, gel-like medium in which the cells and fibers are embedded. Ground substance has several primary functions. It contains a high proportion of water and this accounts for the first of its primary functions—diffusion of nutrients and waste products. A second function of the ground substance is to provide a mechanical barrier against invading bacteria and microorganisms. Connective tissue cells, being part of the reticuloendothelial system, provide the first line of defense against invading organisms. A third function of ground substance is to maintain the so-called "critical interfiber distance." Collagen fibers that approximate one another can potentially adhere together if a certain distance is not maintained between them. The ground substance, which provides some of the tissue volume, can effectively maintain the distance between fibers preventing microadhesions and maintaining extensibility. Ground substance content in connective tissue seems to decrease with age, possibly contributing to a decrease in flexibility with aging.

The primary components of ground substance are glycosaminoglycans (GAGs) and water. Glycosaminoglycans are also referred to as "acid mucopolysaccharides" in the older literature. The two major groups of GAGs are the sulfated and nonsulfated groups. The nonsulfated group, which is primarily hyaluronic acid, acts primarily to bind water. Water makes up approximately 70 percent of the total connective tissue content.[3,4]

Hyaluronic acid, which has long been used to help restore joint function in the veterinary world, has now received Food and Drug Administration approval for use in the injection of human joints. Chondroitin, which is another component of ground substance, is being sold in alternative medicine settings, presumably "to help joint function." The idea of using nonhormonal components of connective tissue to help restore the tissue is an idea that is budding and will have a great impact on the long-term treatment of injured or arthritic joints.

Biosynthesis of Collagen

Collagen synthesis begins in the fibroblast by the absorption of amino acids into the cell. In the rough endoplasmic reticulum of the cell, the amino acids are synthesized into polypeptide chains. From the polypeptide chains, protocollagen (or procollagen), a precursor of collagen is synthesized. Strands of protocollagen are linked in a triple helix in the cell to form strands of tropocollagen. Tropocollagen, which is the molecular unit of collagen, is then passed through the cell membrane into the interstitial spaces. In the extracellular space, tropocollagen strands are linked in series and in parallel in a quarter stagger arrangement to form collagen fibrils (Figures 3–3 and 3–4). Initially, the tropocollagen molecules are hydrostatically attracted to

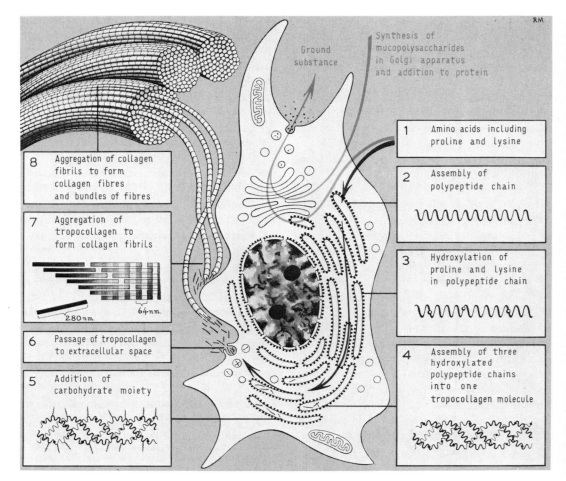

Figure 3–3 A schematic drawing representing the biosynthesis of collagen by fibroblasts. *Source:* Reprinted from *Gray's Anatomy*, ed 35 (p 38) by P. Williams and R. Warwick with permission of W.B. Saunders, © 1973.

each other and form hydrostatic bonds. Eventually, the collagen matures and the weak hydrostatic bonds are converted to stronger covalent bonds.[8]

To review briefly, hydrostatic bonds are those in which polarized molecules or molecules of different polarities are attracted to and weakly bonded to one another. Covalent bonds are bonds in which the two bonding atoms in the respective molecules share an electron. The energy re-

quired to break a covalent bond is much greater than the energy required to break a hydrostatic bond. This accounts for the increasing strength of collagenous tissue during maturati-on. Collagen fibrils eventually band together to form collagen fibers. The configuration of mature collagen can be likened to the structure of common rope. Small strands intertwine to form larger strands; larger strands intertwine to form even larger strands, and so forth (Figure 3–5).

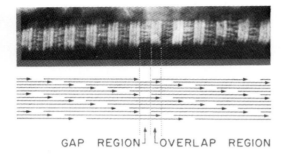

Figure 3–4 Top Electron micrograph showing alternating light and dark regions, and **Bottom** showing the proposed quarter stagger arrangement of collagen fibers. *Source:* Reprinted from *Histology* (p 234) by A.W. Ham and D.H. Cormack with permission of J.B. Lippincott Co., © 1979.

Biomechanics of Connective Tissue

General Characteristics and Definition of Terms

All injuries, whether to bone or connective tissues, are caused by forces acting on these tissues. In order to prevent and treat these injuries, the manual therapist must first have a working knowledge of the basic guiding biomechanical principles that apply to soft tissues. When a force is applied to connective tissues (mechanical stress), the tissues tend to resist any changes in size or shape. Some deformation or change in length can occur, however, as a result of the stress. This deformation is called "strain." Strain is determined by comparing change in length with the normal length. Strain is expressed in deformation per unit length, or percentage change. Tissue strain can be caused by stresses such as a push, pull, twist, tension, compression, or shear. The latter three are common factors in connective tissue injury.[9]

Tension is a pulling force along the length of the tissue. An example of this is in a whiplash injury. The cervical spine is flexed and extended with force. The posterior and anterior ligaments get tightened or stretched and subjected to tension stress.[9,10]

Compression occurs when there is stress applied along the length of a tissue, but the tissue

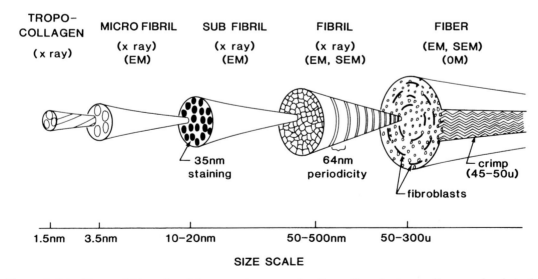

Figure 3–5 Architectural hierarchy of dense regular connective tissue, from the tropocollagen molecule to the collagen fiber. *Source:* Adapted with permission from J. Kastelic, A. Galeski and E. Baer, The multicomposite structure of tendon, *Connective Tissue Research* (1978;6:11–23), Copyright © 1978, Gordon and Breach Science Publishers.

decreases in length and increases in perimeter. In an upright position, compression force is put through the intervertebral discs. The two surfaces become closer to each other as the sides (annulus fibrosis) bulge out under tension.[9,10]

Shearing occurs when one part of a tissue slides over another. This occurs when forces in opposite direction are applied to a tissue. An example of this is L5 sliding forward over S1, leading to a higher incidence of disc herniation at this level.[9,10]

As previously mentioned, when stress is applied to a tissue, deformation occurs. This deformation is called "strain." The strain, or change in length, can be temporary or permanent. A graphic representation of this relationship would appear as a stress/strain curve. Initial change in length requires little force. As more stress is applied to the tissue, the change in length diminishes. In other words, greater amounts of force are required to effect small amounts of change. The early part of the curve, sometimes called *the toe region,* represents the elastic component of connective tissue. This usually represents temporary length changes in the tissue. When the material stretches beyond the elastic range, it reaches a point at which the deformation becomes permanent. This point is called *the elastic limit.* If stress continues, the tissue moves into the *viscous* or plastic range. The tissue is now permanently deformed, but does not rupture. As the imposed stress increases further, the curve reaches its peak at the *yield point.*[9]

Viscoelastic model of connective tissue. This concept can be explained further using a simple engineering model. Connective tissue is sometimes referred to as being *viscoelastic* in nature. It contains both a viscous (permanent) deformation characteristic and an elastic (or temporary) deformation characteristic. The two characteristics combine to give connective tissue its unique qualities.[11-15] This model incorporates a spring (elastic) and a hydraulic cylinder (plastic) linked in series to help depict this deformation quality (Figure 3–6).

The elastic component of connective tissue represents the temporary change in length when subjected to stretch (spring portion of model). The elastic component has a post-stretch recoil in which all length or extensibility gained during stretch or mobilization is lost over a short period of time (Figure 3–7). In the elastic model, the spring recoils when tension or force is removed. The elastic component is not well understood but is believed to be the *slack* taken out of the connective tissue fibers. For example, a regular connective tissue has a loose basket weave configuration of collagen fibers. When a stretch is placed on the tissue, the slack is taken out as the fibers align themselves in the general direction of the stretch (Figure 3–8). When the stretch is removed, the fibers assume their previous orientation and the change in length is lost.

The *viscous* (or plastic) component represents the permanent deformation characteristic of connective tissue. After stretch or mobilization, part of the length or extensibility gained remains even after a period of time (hydraulic cylinder portion of model). There is no post-mobilization recoil in this component (Figure 3–9). In the model, the hydraulic cylinder has been opened and does not close. Presumably, the permanent change results from breaking intermolecular and intramolecular bonds between collagen molecules, fibers, and cross links.

The *viscoelastic* model is then simply the viscous and elastic portions of the model combined and arranged in series (Figure 3–10). After a force is applied to the connective tissue through stretch or mobilization, a net change in length is achieved. Some of the change is quickly lost, while some remains.

The combination of viscous and elastic properties allows for connective tissue to respond by creep and relaxation.[10] Creep occurs when a load is applied to a tissue over a prolonged period of time, as in progressive stretching. This allows a gradual elongation of the tissue. The degree of deformation is more determined by the duration of force applied to the tissue rather than the amount of force. A lesser load over a greater period of time will produce a larger amount of

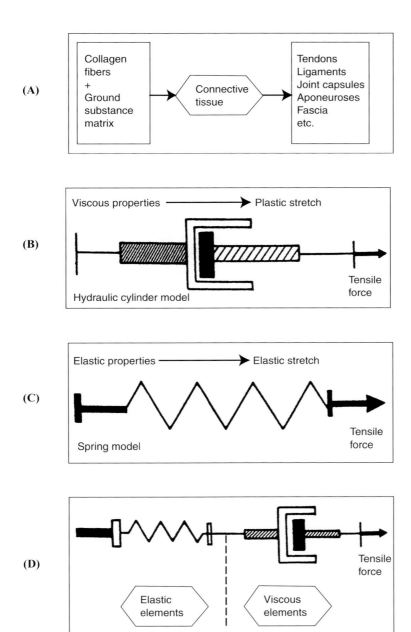

Figure 3–6 **(A)** The primary and secondary organization of connective tissue in the body. **(B)** Schematic representation of a viscous element in material capable of permanent (plastic) deformation. **(C)** Schematic representation of an elastic element in material capable of recoverable (elastic) deformation. **(D)** A simplified model of collagenous tissue. Connective tissue is a viscoelastic material: When stretched, it behaves as if it has both viscous and elastic elements connected in series. *Source:* Reprinted with permission from *The Physician and Sports Medicine*, Vol. 9, No. 12, p. 58, © 1981, McGraw-Hill Companies.

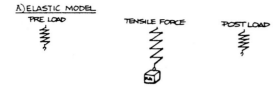

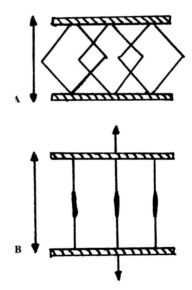

Figure 3–7 Schematic representation of the visco-elastic model of elongation—elastic component in which no permanent elongation occurs after application of tensile force. *Source:* Reprinted from *Myofascial Manipulation: Theory and Clinical Management* (p 4) by A.J. Grodin and R. Cantu with permission of Forum Medicum Inc, © 1989.

creep. An elevation in temperature will cause corresponding increases in creep. Hence, when stretching tight connective tissue, warmed tissue held for a sustained period will be more pliable than cold tissue stretched quickly.[9,10]

If force is applied intermittently, as in progressive stretching, a progressive elongation may be achieved. In Figure 3–11A, strain, or percent elongation, is plotted against time for the purposes of illustrating this phenomenon. Initially, there is a rapid elongation of the tissue, again representing the contribution of the elastic portion of connective tissue. As time passes, less elongation is achieved, representing the contribution of the viscous portion of connective tissue. When the stress is eventually released, the tissue immediately loses some of the previously attained elongation. Again, this phenomenon is consistent with the elastic characteristics of connective tissue. Not all the change in length is lost, however, because the tissue was stretched into the viscous or plastic range.

If the stress is reapplied to the tissue, the curve looks identical, but starts from the new length achieved after the first stretch (Figure 3–11B). Again, the initial elongation is very rapid, but gradually slows as the tissue makes the transition from elasticity to plasticity. When the stress is re-released, another portion of the change in length is lost, and a portion is also retained. With each progressive stretch, the tissue has

some gain in total length that is considered permanent.

This phenomenon is seen often in the clinical setting. In stretching a restricted joint capsule, for example, a certain increase in range of motion may be achieved during a particular treatment session. The patient may return a day or two later with a range of motion greater than the original range, but less than that achieved at the end of the previous treatment. In other words, some range is lost due to the elastic component, and some is retained due to the plastic, or viscous, component.

Although the plastic component represents a permanent elongation, connective tissue is still capable of losing the elongation. The half-life of collagen is 300 to 500 days in mature

Figure 3–8 Diagram showing the weave pattern of collagen, with A and B representing elastic stretch and recoil of collagen fibers. *Source:* Reprinted from Donatelli R. and Owens-Burkhart, H., Effects of Immobilization on the Extensibility of Periarticular Connective Tissue, *Journal of Orthopaedic and Sports Physical Therapy*, Vol. 3, pp. 67–72, with permission of the Orthopaedic and Sports Sections of the American Physical Therapy Association.

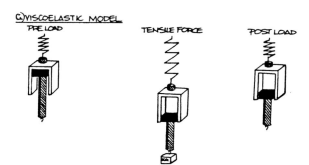

Figure 3–9 Schematic representation of the viscoelastic model of elongation—plastic component in which deformation remains after the application of tensile force. *Source:* Reprinted from *Myofascial Manipulation: Theory and Clinical Management* (p 5) by A.J. Grodin and R. Cantu with permission of Forum Medicum Inc, © 1989.

Figure 3–10 Schematic representation of the viscoelastic model of elongation—some elongation is lost and some is retained after the application of tensile force. *Source:* Reprinted from *Myofascial Manipulation: Theory and Clinical Management* (p 5) by A.J. Grodin and R. Cantu with permission of Forum Medicum Inc, © 1989.

nontraumatized conditions.[16] Over time, new collagen is laid down to replace older collagen. New collagen is laid down according to stresses (or lack of stresses) applied to the tissue. If the tissue is not stressed for long periods of time, it will adaptively shorten as collagen is laid down in the context of the length of the tissues and lack of stresses applied. Wolff's law, which states that "bone adapts to the stresses applied,"[7] can be applied to connective tissue. All connective

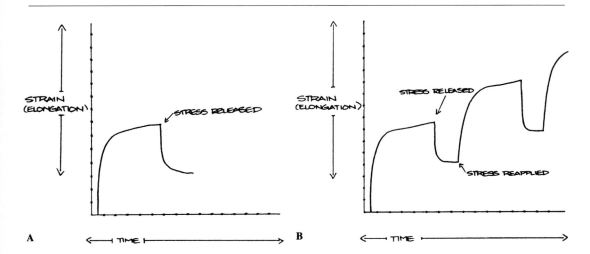

Figure 3–11 (A) Elongation of connective tissue (strain) plotted against time. **(B)** Repeated elongations of connective tissue (strain) plotted against time. *Source:* Reprinted from *Myofascial Manipulation: Theory and Clinical Management* (pp 5–6) by A.F. Grodin and R. Cantu with permission of Forum Medicum Inc, © 1989.

tissue seeks metabolic homeostasis commensurate with the stresses being applied to that particular tissue. Wolff's law, however, applied to connective tissue, has a functional as well as a dysfunctional aspect. Abnormal stresses chronically applied to connective tissues may change the tissue resulting in dysfunction in the tissues and the adjacent structures supported by that tissue (i.e., facet joints, etc.). A clinical example of this phenomenon is the connective tissue band that develops in the patient with spondylolisthesis. Because the spine in this condition cannot withstand the anterior shear forces applied daily, the body responds by laying down connective tissue, in time forming a connective tissue band. Normal stresses, or carefully controlled stresses (i.e., those stresses imparted externally by the clinician in the form of manipulation, or by the patient, in the form of exercises), may positively change the metabolic and physical homeostasis of the tissue. Collagen production is thus less haphazard, more organized, and laid down in a quantity and direction more suited to optimal tissue function. This concept is more fully developed in Chapter 4.

Specific Characteristics

Dense regular connective tissue. Ligaments and tendons are categorized as dense regular connective tissue. Dense parallel arrangement of collagen fibers characterizes dense regular connective tissue (Figure 3–12). The high proportion of collagen to ground substance and the parallel arrangement of the fibers accounts for the high tensile strength and limited extensibility of these tissues. Because of the histologic makeup of these tissues, they are the least responsive to manual work. Because of the compactness and density of collagen fibers and the relatively small proportions of ground substance, the tissue is not highly metabolic, and not very vascular, accounting for the increased healing time required after trauma.

The primary function of tendon is to attach muscle fibers to bone and to transmit forces expended by muscle to the bone with limited elongation, allowing for tension or joint move-

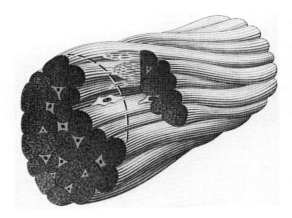

Figure 3–12 Drawing of dense regular connective tissue, showing the parallel arrangement of collagen fibers. *Source:* Reprinted from *Gray's Anatomy*, ed 35 (p 40) by P. Williams and R. Warwick with permission of W.B. Saunders, © 1973.

ment.[17,18] The collagen fibers in tendon have, therefore, been designed in a parallel arrangement to provide the highest unidirectional tensile strength possible. The stress-strain relationship of tendon is similar to that of other connective tissues, with some minor differences. When a tendon is stressed, the toe region (elastic component) of the stress-strain curve is generally smaller due to the parallel arrangement of collagen fibers. This indicates less realignment of fibers than found in other connective tissues during tension. The toe region is generally followed by a moderately linear region with a slightly greater slope, which is indicative of the tendon's greater stiffness. With further tensile deformation, small dips or hitches appear in the curve that possibly represent early tissue microfailure. Finally, with further loading, the tissue fails completely, and the stress-strain curve drops to zero.[17,19]

The primary function of ligament is to check excessive motion in joints and to guide joint motion.[17,18] Ligaments have a less consistent parallel arrangement of collagen fibers than does tendon (Figure 3–13).[20] Under light microscopy, the orientation of the collagen takes on an undulating configuration known as "crimp."[21]

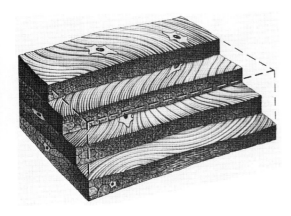

Figure 3–13 Drawing of ligamentous tissue, showing overall parallel arrangement of fibers, but somewhat less parallel than tendon. *Source:* Reprinted from *Gray's Anatomy*, ed 35 (p 40) by P. Williams and R. Warwick with permission of W.B. Saunders, © 1973.

This crimp phenomenon is thought to be responsible for the mildly elastic characteristics of ligament. The ligament functions biomechanically as a spring, until all of the crimp is straightened out and, subsequently, becomes more tensile when the collagen fibers are actually stressed. The ultimate biomechanical result is that ligaments have somewhat less tensile strength per unit area than tendon, but have slightly more yield (Table 3–2).

Dense irregular connective tissue. Dense irregular connective tissue includes, but is not limited to, joint capsules, aponeuroses, penosteum, and fascial sheaths under high degrees of mechanical stress. The major difference between dense irregular and dense regular connective tissue is the orientation of collagen fibers. In dense irregular connective tissue, the collagen fibers are aligned multidirectionally in order to withstand multidirectional stresses (Figure 3–14). The lumbodorsal fascia, for example, has many different attachments, and is pulled in different directions during the spine's normal function.

Loose irregular connective tissue. Loose irregular connective tissue includes, but is not limited to, the superficial and some deep fascia, as well as muscle and nerve sheaths. The supportive framework of the lymph system and the internal organs is also classified as loose irregular connective tissue. Loose irregular connective tissue is generally characterized by a sparse, multidirectional framework of collagen and elastin. Loose irregular connective tissue contains a greater amount of ground substance per unit area than other types of connective tissues. Because of sparse concentrations of collagen in this type of tissue, loose irregular connective tissue is the most elastic and typically has the greatest potential for change when manipulated by external forces.

Table 3–2 Classification of Connective Tissue

Tissue Type	Specific Structures	Characteristics of the Tissue
Dense regular	Ligaments, tendons	Dense, parallel arrangement of collagen fibers; proportionally less ground substance
Dense irregular	Aponeurosis, periosteum, joint capsules, dermis of skin, areas of high mechanical stress	Dense, multidirectional arrangement of collagen fibers; able to resist multidirectional stress
Loose irregular	Superficial fascial sheaths, muscle and nerve sheaths, support sheaths of internal organs	Sparse, multidirectional arrangement of collagen fibers; greater amounts of elastin present

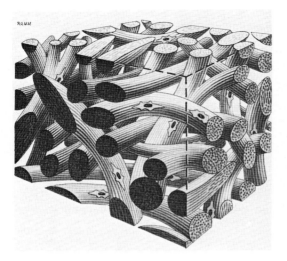

Figure 3–14 Drawing of dense irregular connective tissue, showing the multidirectionality as well as high density of collagen fibers. *Source:* Reprinted from *Gray's Anatomy*, ed 35 (p 40) by P. Williams and R. Warwick with permission of W.B. Saunders, © 1973.

HISTOLOGY AND BIOMECHANICS OF MUSCLE

As previously stated, the myofascial tissues account for the majority of tissue being affected by orthopedic manual therapy. A large portion of the myofascial tissues includes muscle tissue. As with connective tissue, a basic understanding of muscle tissue is also essential for an appropriate empirical understanding of myofascial manipulation. Knowledge of trauma, immobilization, and remobilization of muscle tissue must be built based on the scientific principles that will be outlined as follows. The histology and physiology of muscle tissue alone occupies whole chapters in textbooks. The purpose of this section is to provide a basic overview of muscle histology and how it relates to connective tissue. Much of the knowledge of mammalian skeletal muscle comes from studies of frog skeletal muscle, which is anatomically and histologically similar.

Histology

Muscle is histologically categorized into three types: skeletal, smooth, and cardiac. This section focuses primarily on skeletal muscle, which in turn will provide a basis for understanding cardiac and smooth muscle types. Skeletal, or striated muscle, is so named because of its striated or banded appearance under light microscopy. The striations reflect the functional contractile unit of the muscle called the *sarcomere*. Muscle is also functionally characterized by fiber type based on speed of contraction or relaxation, biochemistry and metabolism, and in circulation.

Mechanism of Growth in Skeletal Muscle

The total number of actual muscle fibers in a muscle is reached sometime before birth. Longitudinal growth in a muscle is accomplished in early years by an increase in the length of the individual sarcomeres and by addition of sarcomeres. Increases in diameter are accomplished by the addition of *myofilaments* in parallel arrangement. Likewise, the muscle shortens by losing sarcomeres and decreases in diameter by losing myofilaments. With prolonged disuse, the muscle fibers degenerate and the tissue is replaced with less metabolically active connective tissue. Human skeletal muscle, however, does have some limited regeneration potential. Satellite cells, which are believed to be a persisting version of the prenatal myotubes found inside basement membranes, can become activated to produce a limited amount of new muscle fibers. The number of new fibers that can be produced, however, cannot compensate for the amount lost during major muscle trauma or degeneration.

Cellular and Histological Organization of Skeletal Muscle

The contractile proteins of striated muscle are actin and myosin. The actin and myosin interact in a ratchet-type manner to shorten the muscle (Figure 3–15). Actin and myosin filaments are contained in the functional contractile unit of muscle called the sarcomere. The

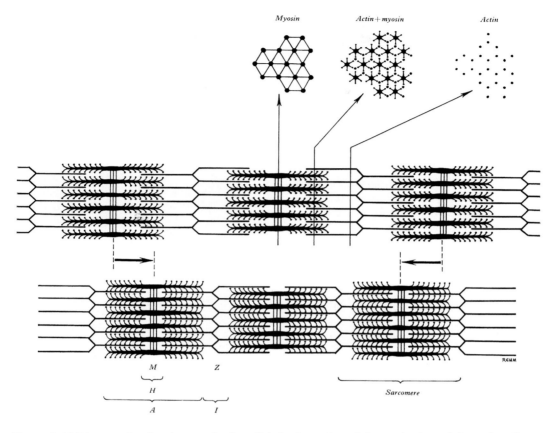

Figure 3–15 Diagram showing the organization of skeletal muscle and the mechanism of shortening. *Source:* Reprinted from *Gray's Anatomy*, ed 35 (p 479) by P. Williams and R. Warwick with permission of W.B. Saunders, © 1973.

transverse alignment of sarcomeres in adjacent myofilaments gives this tissue the striated appearance. The striations result from a series of bands (Z, A, I bands), which reflect components of the sarcomere. The distance between two Z bands reflects the length of the sarcomere and will vary depending on the contractile state of the muscle. The A band, which represents myosin molecules, does not change in length during contraction, whereas the I band, which represents areas where actin does not overlap myosin, changes depending on the contractile state of the muscle.

Sarcomeres are arranged in series to form cylindrical organelles called myofilaments. Myofilaments are arranged in bundles and are contained in the *myofibril*, which is the muscle's cellular unit. Myofibrils are multinucleated cells that also contain mitochondria, lysosomes, ribosomes, and glycogen. Myofibrils are grouped together into bundles called *fasciculi*. Loose connective tissue fills the area between myofibrils and is called endomysium. A loose connective tissue sheath also surrounds the muscle fasciculus and is called the perimysium. Finally, fasciculi are grouped together to form individual muscles (Figure 3–16). The loose connective tissue sheath that envelops the muscle is called the epimysium.

Biomechanics of Muscle

The connective tissues of skeletal muscle have important roles in the optimal function of muscle. These connective tissues provide a certain amount of coherence in the muscle while allowing an appropriate degree of mechanical freedom. The connective tissue layers also serve to carry the blood supply to the tissue and ramify to form a rich capillary network in the muscle fiber.[21] They also allow the penetration of nerves along with this blood supply to allow for diffusion of nutrients and ions as necessary for muscular metabolism and excitation.[3] The endomysium is particularly significant in these roles, since it most closely approximates the individual muscle fibers.

Muscle Fiber Types

Human muscle is a mixture of Type I and Type II fibers. There is variability in the relative percentages of each type between individuals. Within an individual, there is a correlation between muscle function and fiber composition.[21] Muscles are generally categorized according to the predominant fiber type present throughout the muscle.

The following fiber type classification is currently the most widely used.[22] Fibers are classified as Type I, IIa, IIb, or IIm (Table 3–3). Type I fibers are slow twitch fibers that have the slowest contraction times. They are also the lowest in glycogen stores, but have the richest concentration of mitochondria and myoglobin. Because of these characteristics, Type I fibers are the slowest to fatigue. The postural muscles of the body have a predominance of Type I fibers. Type IIa fibers (also called fast twitch/ioxidative or fast red fibers) are intermediate fibers that have a faster contraction time than Type I fibers while remaining moderately fatigue resistant. A high concentration of myoglobin and mitochondria is still present in these

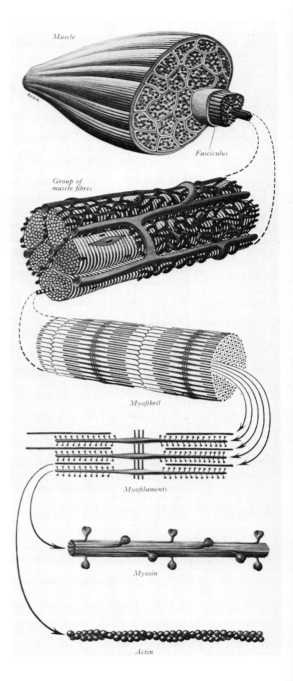

Figure 3–16 Diagram showing architectural hierarchy of muscle tissue. *Source:* Reprinted from *Gray's Anatomy*, ed 35 (p 481) by P. Williams and R. Warwick with permission of W.B. Saunders, © 1973.

Table 3–3 Classification of Muscle Fiber Types

Fiber Type	Functional Classification	Metabolic Characteristics	Functional Characteristics
Type I	Slow twitch	High concentrations of myoglobin, increased numbers of mitochondria, low content of glycogen, oxidative metabolism	Slow contraction times, fatigue resistant
Type IIa	Fast twitch/oxidative (fast red)	Moderately high concentrations of myoglobin, increased numbers of mitochondria, glycolytic/oxidative (mixed) metabolism	Faster contraction times than type 1, less fatigue resistant
Type IIb	Fast twitch/glycolytic (fast white)	High glycogen content, glycolytic metabolism, decreased numbers of mitochondria	Fast contraction times, fatigues easily
Type IIm	Superfast	Contains unique myosin configuration, high glycogen content, glycolytic metabolism	Very fast contraction times

fibers. Type IIb muscle fibers (also called fast twitch/glycolytic or fast white fibers) have faster contraction times and rely more on glycolytic pathways for energy metabolism. Alternately, Type IIb fibers have a lower concentration of myoglobin and mitochondria and are not fatigue resistant. Finally, a superfast fiber, termed IIm, has been identified in mammalian muscle tissue, including human muscle tissue. This type of fiber is found primarily in the jaw muscles and contains a unique myosin that distinguishes it from Types I and II fibers.[23] Muscles with a greater percentage of Type II fibers, those which cross two joints and those working eccentrically, are much more susceptible to strain injuries. The most common site of those injuries is at the musculoskeletal junction.[21]

HISTOLOGY AND BIOMECHANICS OF JUNCTIONAL ZONES

The junctional zones in the myofascial tissues include the myotendinous junction and the ligament, tendon, and joint capsule insertions to bone. Early studies indicate that although injury can occur in any portion of the myofascial tissues, injury to the junctional zones is quite common.[24] Numerous recent stress-strain studies indicate that most tissue failures occur at or near the myotendinous junction.[25–30] Myofascial restrictions will commonly be found in the areas of the junctional zones due to the frequency of injury to these areas, and the clinician should be aware of these areas in myofascial evaluation. A basic understanding of the histology and biomechanics of junctional zones is, therefore, preliminary to a study of their histopathology and to an empirical understanding of myofascial evaluation and treatment.

Histology of Myotendinous Junction

The attachment of the muscle is generally through tendon. The muscle belly attaches to tendon at the musculotendinous junction on each side of the belly. These musculotendinous junctions are highly specialized areas.

Several histological differences occur in the transitional area between muscle fibers and tendon that give it unique functional characteristics. First, the cell membrane forms a continuous interface between intercellular components of muscle fibers and extracellular components of connective tissue. The cell membrane at this junction becomes highly folded or convoluted allowing the contractile intercellular components to interdigitate with the extracellular components.[31–38] The folding of the cell membrane increases the surface area, thereby reducing the stress per unit area on the membrane. The folds hold the membrane at a low angle in relation to the forces coming from the muscle fibers, placing the membrane primarily under shear forces. If the folds did not exist, the junctional membrane would experience vector forces at right angles to the membrane surfaces. This would create a tensile load at the junction. Studies indicate, however, that cell membranes are highly resistant to shear forces that would increase their surface area.[39] The design of the folds allows for much higher force transmission before tissue rupture.

Finally, the membranous folds increase the potential adhesive area in the musculotendinous junction.[40,41] This also decreases the load per unit area being transmitted from the muscle. Interestingly, muscles with predominantly fast twitch muscle fibers have an increased folding of the junctional membranes. This phenomenon is probably related to the fact that higher forces are developed in fast twitch muscles than in slow twitch muscles, and greater cumulative tensile strength is required to sustain and transmit such forces.

Another significant histological characteristic of the myotendinous junction is decreased sarcomere length and extensibility.[42,43] This characteristic results in the myotendinous junction first being loaded by terminal sarcomeres and subsequently being fully loaded by the rest of the sarcomeres in the muscle belly. More significantly, the decreased extensibility of the terminal sarcomeres also makes the tissue in this area more vulnerable to tearing, as evidenced by the frequency of injury in the experimental models.[42,43] The clinical implications are discussed further in Chapter 4.

Biomechanics of the Myotendinous Junction

As previously mentioned, the intercellular contractile units must ultimately be coupled with the collagen fibers of the tendon for transfer of forces to take place (Figure 3–17). This is accomplished architecturally in the following manner.

Thin myofilaments, believed to be derivatives of actin, attach from the terminal Z disks of the

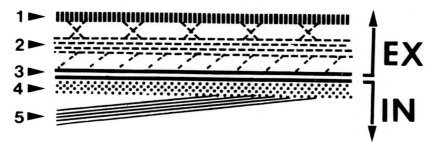

Figure 3–17 Schematic drawing of the structures involved in force transmission between tendon and contractile proteins. Extracellular components (EX) include tendon collagen fibers (1) and basement membrane (2). The junctional plasma membrane (3) separates extracellular (EX) and intracellular (IN) force-transmitting structures. Within the cell, thin actin filaments (5) are attached to the cell membrane by dense, subsarcolemmal material (4). *Source:* Reprinted from *Injury and Repair of the Musculoskeletal Soft Tissues* (p 184) by SL.-Y. Woo and J.A. Buckwalter with permission of the American Academy of Orthopaedic Surgeons, © 1987.

myofibrils to a thickened cell area of the inner cell membrane called the subsarcolemma. The contractile proteins of the muscle sarcomeres, therefore, have an attachment to the cell membrane. The outer portion of the cell membrane is similarly attached to a basement membrane that runs parallel to the cell membrane. The basement membrane contains type IV collagen and high molecular weight glycoproteins. The basement membrane is then attached to collagen fibers of the tendon.[32–35,37,38]

As can be seen in Figure 3–17, all of the components of the myotendinous junction are coupled in a parallel arrangement, rather than in series. As previously mentioned, the cell membrane can accommodate shear forces more optimally than tensile forces, and the architecture of the myotendinous junction reflects this efficiency.

Connective Tissue Insertion to Bone

The insertions of tendons, ligaments, and joint capsules to bone vary somewhat in their histologic architecture. As with the myotendinous junction, the architecture is designed to dissipate tensile forces and minimize stress concentrations. Despite their architectural design, these junctions are common sites of injury and remain areas of weakness during loading. As with the other areas examined in this chapter, a basic review of the histology and biomechanics of these junctions is necessary to understand their response to trauma and pathology.

Within an area of 1 millimeter, the connective tissue is transformed into hard tissue (Figures 3–18A and B). Two types of insertions are identified in the literature: direct and indirect. Direct insertions have four distinct histological zones that represent the transition of the tissues from soft connective tissues to bone.[44]

Zone 1 consists of the actual tendon or ligament. The histology of this zone does not differ much from the histology of ordinary tendon, ligament, or capsule. Collagen fibers are found here embedded in the matrix or ground substance, as are fibroblasts. Zone 2 consists of fibrocartilage. The cells in this region resemble chondroblasts or chondrocytes. Zone 3 consists of mineralized fibrocartilage, where mineral deposits are found around collagen fibrils. Finally, zone 4 consists of bone, where the collagen fibrils merge with the fibrils of the bone matrix.

Indirect insertions do not have specifically defined zones as do the direct insertions. The connective tissue fibers tend to blend more with the periosteum, which in turn attaches to bone. These transitional fibers are sometimes referred to as Sharpey's fibers.[44] These fibers are described as originating in the periosteum and perforating the underlying bone, anchoring the periosteum to underlying bone.[18] No fibrocartilage is seen in indirect insertions.[45,46]

A common feature of the two insertional types is the presence of superficial and deep fibers. The superficial fibers generally attach to periosteum, which in turn attaches to bone. The deep fibers insert into bone or by way of fibrocartilage. The main difference is that the direct insertion has a fibrocartilaginous transitional zone, while the indirect insertions do not. Another commonality is that the junctional zones of ligament, tendon, and capsule are relatively avascular compared with the tissue on either side of the zone.[47,48]

The attachment sites of ligament, tendon, and joint capsule to bone also vary in their biomechanics because of differences in the forces imparted by these tissues. Obviously, the tendon-bone junction will have greater forces because of the forces generated by muscle, whereas the ligament and joint capsule-bone junction will have lesser forces. The resiliency of the tendon-bone junction was demonstrated by Noyes and associates.[49] Several samples of patellar tendon were analyzed to determine stress-strain characteristics of the tendon proper, the entire bone-tendon-bone unit, and the actual attachment site. The attachment sites undergo more significant strain (elongation) before receiving significant stresses, indicating that strains in this region are greater than any other region. This allows for more force dissipation at this region, but also makes this region more vulnerable.

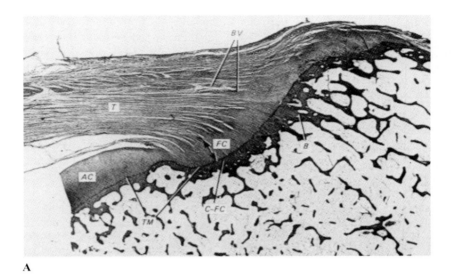

A

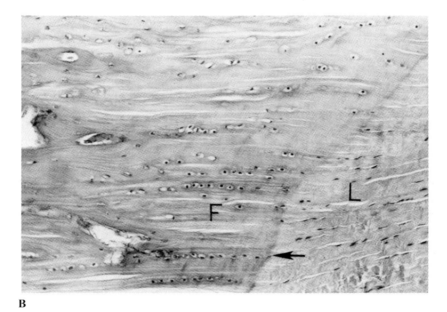

B

Figure 3–18 (**A**) Direct insertions. The four distinct zones seen in the supraspinatus insertion. The four zones are tendon (T), uncalcified fibrocartilage (FC), and bone (B). *Source:* Reprinted with permission from M. Benjamin, E.J. Evans, et al., The Histology of Tendon Attachments to Bone in Man, *Journal of Anatomy*, No. 149, pp. 89–100, © 1986, Cambridge University Press. (**B**) Femoral insertion of rabbit MCL. The deep fibers of the ligament (L) pass into bone through the fibrocartilage (F). The arrow indicates the line of calcification. *Source:* Reprinted with permission from SL.-Y. Woo, M.A. Gomez et al., The Biomechanical and Morphological Changes in the Medial Collateral Ligament of the Rabbit after Immobilization and Remobilization, *Journal of Bone & Joint Surgery*, Figure 6-A, Vol. 69A, p. 1207, © 1987, Journal of Bone & Joint Surgery.

CONCLUSION

The information covered in this chapter was primarily of a basic science nature. Although somewhat removed from the clinical realm, a thorough understanding of basic anatomy and biomechanics is necessary for the manual physical therapist to be successful in the treatment of the myofascial tissues. This understanding will allow the therapist to set realistic goals for manual treatment. In this day and age, where various types of practitioners are competing for patients, and reimbursement by insurance companies is decreasing, it is essential for our profession to establish credibility in what we do. Art and science must be carefully balanced as the profession forges ahead, especially in the area of myofascial manipulation.

REFERENCES

1. Dicke E, Schliack H, Wolff A. *A Manual of Reflexive Therapy of the Connective Tissue*. Scarsdale, NY: Sidney S. Simon Publishers; 1978.

2. Ham AW, Cormack DH. *Histology*. Philadelphia: JB Lippincott: 1979:210–259.

3. Warwick R, Williams PL. *Gray's Anatomy*. 3rd ed (Br). Philadelphia: WB Saunders; 1973:32–41, 480–42.

4. Copenhaver WM, Bunge RP, Bunge MB. *Bailey's Textbook of Histology*. Baltimore, MD: Williams & Wilkins; 1971.

5. Geneser F. *Textbook of Histology*. Philadelphia: Lea & Febiger; 1986.

6. Fielding JW, Burstein AH, et al. The nuchal ligament. *Spine*. 1976;1:3.

7. Nachemson AL, Evans JH. Some mechanical properties of the third human lumbar interlaminar ligament (ligamentum flavum). *J Biomech*. 1968;l:211.

8. Cummings G, Crutchfield CA, Barnes MR. *Soft Tissue Changes in Contractures*. Atlanta, GA: Stokesville Publishing; 1985.

9. Norris C. *Sports Injuries: Diagnosis and Management*. Oxford: Butterworth Heinmann LTD; 1993.

10. Bernhardt D. *Sports Physical Therapy*. New York: Churchill Livingstone; 1986.

11. Sapega AA, Quedenfeld TC. Biophysical factors in range of motion exercise. *Physician Sports Med*. 1981; 9:57–65.

12. Warren CG, Lehmann JF, et al. Heat and stretch procedures: An evaluation using rat tail tendon. *Arch Phys Med Rehabil*. 1976;57:122-126.

13. Woo SL-Y, Ritter D, et al. The biomechanical and biochemical properties of swine tendons: Long-term effects of exercise on the digital extensors. *Connect Tissue Res*. 1980;7:177–183.

14. Fung YCB. Elasticity of soft tissues in simple elongation. *Am J Physiol*. 1967;213:1532–1545.

15. Hooley CJ, McCrum NG, et al. The viscoelastic deformation of tendon. *J Biomech*. 1980;13:521–528.

16. Neuberger A, Slack H. The metabolism of collagen from liver, bones, skin and tendon in normal rat. *Biochem J*. 1953;53:47–52.

17. Frankel VH, Nordin M. *Basic Biomechanics of the Skeletal System*. Philadelphia: Lea & Febiger; 1980:56, 87–110.

18. Woo SL-Y, Buckwalter JA. *Injury and Repair of the Musculoskeletal Soft Tissues*. Savannah, GA: American Academy of Orthopaedic Surgeons Symposium; 1987.

19. Viidik A. Tensile strength properties of Achilles tendon systems in trained and untrained rabbits. *Acta Orthop Scand*. 1969;40:261–272.

20. Kennedy JC, Hawkins RJ, et al. Tension studies of human knee ligaments. Yield point, ultimate failure, and disruption of the cruciate and tibial collateral ligaments. *J Bone Joint Surg*. 1976;58:A350–A355.

21. Barlow Y, Willoughby S. Pathophysiology of Soft Tissue Repair. *Br Med Bull*. 1992;48(3),698–711.

22. Gauthier GF. Skeletal muscle fiber types. In: Engel AG, Banker BQ, eds. *Myology*. New York: McGraw-Hill; 1986;1:255–284.

23. Rowlerson A, Pope B, et al. A novel myosin present in cat jaw closing muscles. *J Muscle Res Cell Motil*. 1981;2:415–438.

24. McMaster PE. Tendon and muscle ruptures: Clinical and experimental studies on the causes and location of subcutaneous ruptures. *J Bone Joint Surg*. 1933;11.5: 705–722.

25. Almekinders LC, Garrett WE Jr, et al. Pathophysiologic response to muscle tears in stretching injuries. *Trans Orthop Res Soc*. 1984;9:384.

26. Almekinders LC, Garrett WE Jr, et al. Histopathology of muscle tears in stretching injuries. *Trans Orthop Res Soc*. 1984;9:306.

27. Garrett WE Jr, Almekinders LC, et al. Biomechanics of muscle tears in stretching injuries. *Trans Orthop Res Soc*. 1984;9:384.

28. Garrett WE Jr, Nikalaou PK, et al. The effect of muscle architecture an the biomechanical failure properties of skeletal muscle under passive extension. *Am J Sports Med* 1988;16:7–12.

29. Nikolaou PK, Macdonald BL, et al. Biomechanical and histological evaluation of muscle after controlled strain injury. *Am J Sports Med*. 1987;15:9–14.

30. Garrett WE Jr, Rich FR, et al. Computed tomography of hamstring muscle strains. *Med Sci Sports Exerc* 1989; 21:506–514.

31. Gelher D, Moore DH, et al. Observations of the myatendan junction in mammalian skeletal muscle. *Z Zellforsch Mikrosk Anat*. 1960;2:325–336.

32. Mackay B, Harrop TJ, et al. The fine structure of the muscle tendon junction in the rat. *Acta Anat*. 1969;73: 588–604.

33. Tidhall JG, Daniel TL. Myotendinous junctions of tonic muscle cells: structure and loading. *Cell Tissue Res*. 1986;245:315–322.

34. Eisenberg BR, Milton RL. Muscle fiber termination at the tendon in the frog's sartorius: A stereological study. *Am J Anat*. 1984;171:273–284.

35. Tidball JG. The geometry of actin filament membrane associations can modify adhesive strength of the myotendinous junction. *Cell Motil*. 1983;3:439–447.

36. Trotter JA. Hsi K, et al. A morphometric analysis of the muscle-tendon junction. *Anat Rec*. 1985;213:26–32.

37. Mair WGP, Tame FMS. The ultrastructure of the adult and developing human myotendinous junction. *Acta Neuropathol*. I 972;21:239–252.

38. Trotter JA, Eberhard S, et al. Structural connections of the muscle-tendon junction. *Cell Motil*. 1983;3:431–438.

39. Evans EA, Hochmuth RM. Mechanochemical properties of membranes. *Curr Top Membranes Transport*. l978;l0:1–64.

40. Bikerman JJ. Stresses in proper adhints. In: *The Science of Adhesive Joints*. New York: Academic Press; 1968:192–263.

41. Lubkin JL. The theory of adhesive scarf joints. *J Appl Mech*. 1957;24:255–260.

42. Gordon AM, Huxley AF, et al. Tension development in highly stretched vertebrate muscle fibers. *J Physiol*. 1966;184:143–169.

43. Huxley AF, Peachy LD. The maximal length for contraction in vertebrate striated muscle. *J Physiol*. 1961; 156:150–165.

44. Cooper RR, Misol S. Tendon and ligament insertion: A light and electron microscopic study. *J Bone Joint Surg*. 1970;52:A1–A21.

45. Woo SL-Y, Gamez MA, et al. The biomechanical and morphological changes in the medial collateral ligament of the rabbit after immobilization and remobilization. *J Bone Joint Surg*. 1987;69:A1200–A1211.

46. Benjamin M, Evans EJ, et al. The histology of tendon attachments to bone in man. *J Anat*. 1986;149:89–l00.

47. Scapinelli R. Studies on the vasculature of the human knee joint. *Acta Anat*. 1968;70:305–331.

48. Amoczky SP, Rubin RM, et al. Microvasculature of the cruciate ligaments and its response to injury: An experimental study in dogs. *J Bone Joint Surg*. 1979;61: A1221–A1229.

49. Noyes FR, DeLucas JL, et al. Biomechanics of anterior cruciate ligament failure: An analysis of strain-rate sensitivity and mechanisms of failure in primates. *J Bone Joint Surg*. 1974;56:A236–A243.

CHAPTER 4

Histopathology of Myofascia and Physiology of Myofascial Manipulation

Deborah Cobb, Robert I. Cantu, and Alan J. Grodin

HISTOPATHOLOGY OF MYOFASCIA

The basis of all treatment techniques lies in understanding the basic processes of soft tissue healing. In the previous chapter, the normal histology and biomechanics of myofascial tissues were discussed. With that groundwork laid, this chapter will now address the histopathology and pathomechanics of those same tissues. A review of classic as well as recent literature will be used to provide an understanding of scar formation after trauma as well as how myofascial tissues can be affected by immobilization and remobilization. With an awareness of the changes that occur in the myofascial tissues under dysfunctional conditions, a manual therapist can then set realistic treatment goals and choose the most appropriate treatment techniques to accomplish them. The intuitive aspects of myofascial manipulation must always be balanced by a solid understanding of tissues and their response to dysfunction.

Pathophysiology of Soft Tissue Repair

A *wound* by its most basic definition is a disruption of unity. Because vertebrates lack the ability to regenerate exact duplicates of injured parts, response to injury comes in the form of repair through granulation scar tissue. The scar formation process is not a cyclic but a linear process with a sequence of recurring stages. The literature varies as to whether there are three or four distinct phases a wound passes through.[1-4] This chapter will divide the scar process into four distinct phases: (1) the inflammatory phase; (2) the granulation phase; (3) the fibroplastic phase; and (4) the maturation phase.[5,6] Time tables for the beginning and end of each phase must be understood as general guidelines. Different tissues heal at different rates, and within one wound itself areas in various phases of healing may be seen.[1] The changes may also be affected by the age and fitness level of an individual.[7]

Inflammation, a normal prerequisite to healing, is the first phase seen after a trauma. This phase begins immediately and may last 24 to 48 hours. Injury causes chemical and mechanical changes leading to alteration in blood flow. This in turn leads to the cardinal signs of inflammation: heat, redness, swelling, and pain. The inflammatory response to injury is the same regardless of the injuring agent or the location of the injury.[8] Whole blood poured directly into a wound will coagulate and temporarily seal off the injured vessels and lymphatic channels. This traumatic exudate acts to temporarily seal the wound. Histamine is released by the injured tissues; resulting in vasodilatation and the appearance of a reddened, hot, and swollen region.[1] Prostaglandins, formed from cell membrane phospholipids when cell damage occurs, are responsible for pain production.[2] Phagocytosis then occurs to prevent infection in the wound and prepare the wound for healing. Phagocytosis is initiated by short-lived polymorphonuclear leukocytes that first attach to bacteria and then dissolve and digest them. Shortly after, macrophages appear to continue the phagocytic process and to begin influencing scar production.[9] Its role in recruiting fibroblasts is significantly related to the final amount of scar produced.[1] At this point, movement in this area would be disadvantageous and could lead to further tissue and/or clot disruption. Modalities aimed at decreasing inflammation, proper positioning, and appropriate anti-inflammatory medications are of the most value at this point (Figure 4–1).

The *granulation* phase begins when the macrophages and histiocytes debride the area. The granulation stage is so named because of the appearance of capillary buds that microscopically look like granules. Healing cannot proceed further unless this increased connective tissue vascularity can meet the metabolic demands of the healing tissues. Immobilization is essential during this phase to permit vascular regrowth and prevent further microhemorrhages

and tissue breakdown.[1,10] Heat application at this point may cause increased bleeding in the fragile healing tissues.[11]

Rebuilding of tissue begins with the *fibroplastic* phase. Proliferation of fibroblasts and accelerated collagen synthesis now occur. As the fibroblasts proliferate, new collagen is laid down in a disorganized manner in the area of the wound. Strength of the wound is determined not by the amount of collagen laid down but by the bonding of the collagen filaments or *crosslinks* (Figure 4–2).[3] The cross-linking allows for early controlled movement without disruption of the wound. Controlled movement will cause the fibrils to align lengthwise along the line of stress of the healing structure.[12] Because vascularity remains high during this phase, the immature scar still has a characteristic pink coloring. Wound closure usually occurs at this stage, and the time frame varies depending on the vascularity and metabolic rate of the tissue. In tissues with high metabolic activity (muscles, skin, etc.), wound closure occurs in 5 to 8 days. In tissues with lower metabolic activity (ligament and tendon), wound closure occurs in 3 to 5 weeks.[6] During this phase, gentle handling of the wound is essential. Gentle manual therapy

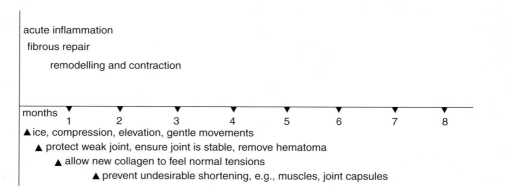

Figure 4–1 Encouraging favorable healing conditions. *Source:* Reprinted with permission from P. Evans, The Healing Process at the Cellular Level, *Physiotherapy*, Vol. 66, No. 8, pp. 256–259, © 1980, Physiotherapy Canada, and G. Hunter, Specific Soft Tissue Mobilization in the Treatment of Soft Tissue Lesions, *Physiotherapy*, Vol. 80, No. 1, pp. 15–21, © 1994, Physiotherapy Canada.

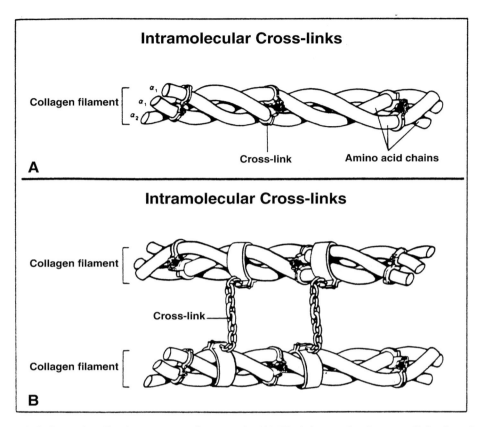

Figure 4–2 Collagen bonding increases tensile strength: (**A**) Weak intramolecular cross-links form between amino acid chains within one collagen filament. (**B**) Stronger intermolecular cross-links form from one collagen filament to another. *Source:* Reprinted from Hardy, A., Biology of Scar Tissue, *Physical Therapy*, Dec. 1989, Vol. 69, No. 12, with permission of the American Physical Therapy Association.

techniques may be appropriate at this time. Soft tissue mobilization designed to break up scar tissue will inflame the wound, leading to further deposition of collagen.[5,6]

The final stage of scar formation is the maturation or remodeling phase. This stage may last from 3 weeks to 12 months.[13] During this phase, collagen must change in order to reach maximum function. A reduction in wound size, a realignment of collagen fibers, and an increase in the strength of the scar are all characteristic of this phase. Arem and Madden[12] confirmed that a physical change in scar length could be achieved through the application of low load,

long duration stress during this phase. During this time, the scar tissue is responsive to manual therapy but the progress will be somewhat slowed. Without controlled stress or mobilization during this phase, however, tensile strength of the scar will not improve and optimal function will be diminished.

Cycle of Fibrosis and Decreasing Mobility in Connective Tissue

The fibrotic process is histologically distinct from the scar formation process. The fibrotic process in connective tissue is a "homogenous" process involving an entire tissue area or the entire tissue "fabric," and does not have clearcut stages as does the scar tissue formation process. The fibrotic process is cyclical in nature, whereas the scar formation process is a linear process that has a distinct end. The fibrotic process in connective tissue can continue as long as the irritant is present.

The fibrotic process is generally initiated by the production of an irritant, possibly traumatic exudates from nearby acutely inflamed traumatized tissue or a low-grade irritation/inflammation of the tissue. The low-grade irritation may be caused by arthrokinematic dysfunction, poor posture, overuse, habit patterns, or structural or movement imbalances. A rotator cuff irritation, for example, may be caused by a poor tennis service, poor sleeping postures, occupational overuse syndromes, and other causes. The mechanical irritant produces a low-grade inflammation, which then starts the process. With an inflammatory response, macrophages are activated to clean and debride the area. In-

flammatory exudates, along with damaged collagen and other waste products, are carried away. The increased metabolic activity in the area stimulates the body to increase the area's vascularity. With increased vascularity and debridement of damaged collagen (from microtrauma), fibroblasts are activated to replace lost collagen. Since the inflammatory process is generally painful, the joint is not being moved in proper fashion. The collagen begins to be laid down in more haphazard arrangement since adequate stress is not being placed on the tissue, and cross-linking with other preexisting collagen fibers begins. At one point, myofibroblasts appear in similar fashion as in the scar process. The myofibroblasts, which contain significant amounts of actin and myosin in the cytoplasm, anchor to adjacent collagen fibers and contract, shrinking the tissue. The tissue shrinkage results in further dysfunctional movement, which, in turn, creates more mechanical stresses and more chronic irritant (Figure 4–3). As long as an irritant is present, the cycle continues.

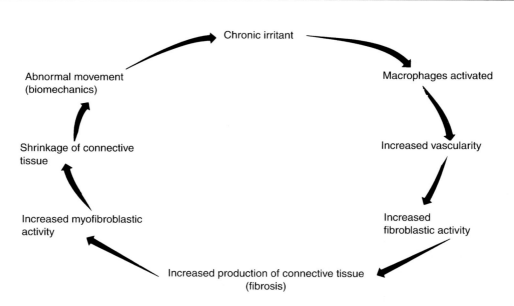

Figure 4–3 Cycle of fibrosis and decreasing mobility.

Response of Myofascial Tissue to Immobilization

Connective tissue has a characteristic histological and biomechanical response to immobilization. Most of the currently available research, however, focuses on animal studies in which an area of the body is immobilized for a period of time, after which the connective tissue is histologically and biomechanically analyzed. Several factors must be considered before applying the results of these studies to the general rehabilitative population. The first is that these are animal studies, the results of which should be cautiously applied to the general human population. Second, and of greater clinical importance, many of the studies that are discussed in this chapter deal with the response of "normal," or nontraumatized, connective tissue to immobilization, and do not necessarily address the responses of traumatized and/or scar tissue. In the general orthopedic setting, connective tissue that has been immobilized has also been traumatized. Trauma does affect the histology and biomechanics of the healing of connective tissue. Also brought into the picture is the process of scar formation, and the effects of immobilization on the developing scar tissue. All of these clinical scenarios are addressed in detail because the response of normal connective tissues to immobilization provides a basis for understanding traumatized conditions.

Nontraumatized Connective Tissue

When a nontraumatized joint is subjected to immobilization, stress-deprived connective cells exhibit changes within 4 to 10 days.[14,15] Intrinsic changes to periarticular connective tissues begin to limit mobility. Much of the early animal studies on immobilized connective tissue were performed by Amiel, Akeson, Woo and their associates.[16–21] In studies utilizing primarily knee joints, laboratory animals were immobilized by internal fixation for periods from 2 to 9 weeks. A pin was placed from the proximal one-third of the femur to the distal one-third of the tibia to presumably avoid traumatizing the knee joint. The animals were then sacrificed at various times of immobilization and the periarticular tissues were analyzed microscopically, histochemically, and biomechanically. From a microscopic standpoint, the authors found fibrofatty infiltrate, especially in the capsular folds and recesses. The longer the immobilization, the greater amount of infiltrate found, along with a change in the infiltrate's appearance, which became more fibrotic. This created macroscopic adhesions in the recesses and capsular folds.

Histological and histochemical analyses showed several significant changes, the primary one being a significant loss in ground substance, with no significant collagen loss. The primary components of lost ground substance were the glycosaminoglycans and water. The authors reported a 30 percent to 40 percent loss in both sulfated and nonsulfated groups. Since the primary purpose of the nonsulfated group (hyaluronic acid) is to bind water, the water loss is easily explained.

As noted in the previous chapter, one of the primary purposes of the ground substance is to lubricate the area between adjacent collagen fibers. Collagen fiber lubrication is associated with the maintenance of the so-called critical interfiber distance. This is the distance that must be maintained between collagen fibers to allow them to glide smoothly and to prevent microadhesions between fibers. When the critical interfiber distance is not maintained, the collagen fibers approximate and eventually become cross-linked by newly synthesized collagen. Also, because collagen fibers are laid down according to the stresses (or lack of stresses) applied, collagen in immobile connective tissue is arranged haphazardly.[22] The newly synthesized collagen then binds adjacent collagen fibers, decreasing the extensibility of the tissue (Figures 4–4 and 4–5).

Several factors explain why significant amounts of ground substance are lost, yet collagen is not. First, the half-life of nontraumatized collagen is 300 to 500 days whereas the half-life of ground substance is 1.7 to 7 days.[23–25] Also, with immobilization times of less than 12 weeks,

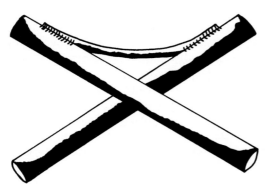

Figure 4–4 Drawing showing the laying down of newly synthesized collagen, forming cross-links onto existing collagen fibers. These cross-links are believed to be responsible for decreased extensibility in immobilized connective tissue. *Source:* Reprinted from Donatelli, R. and Owens-Burkhart, H., Effects of Immobilization on the Extensibility of Periarticular Connective Tissue, *Journal of Orthopaedic and Sports Physical Therapy*, Vol. 3, pp. 67–72, with permission of the Orthopaedic and Sports Sections of the American Physical Therapy Association.

collagen synthesis occurs at the same rate as collagen degradation. After 12 weeks, however, the rate of collagen degradation exceeds the rate

of synthesis, and net amounts of collagen are lost.[26]

Biomechanical analyses indicated that ten times the torque required to move a normal joint was required to move the immobilized joints. After several repetitions, the amount of torque required to move the immobilized joint was reduced to three times that of a normal joint. The biomechanical implication is that fibrofatty macroadhesions and microscopic adhesions in the form of increased collagen cross-linking contributed to the decreased extensibility of the connective tissue.[16–21]

Schollmeier et al immobilized the forelimbs of 10 beagles for 12 weeks. At the end of that time, the passive range of motion of the glenohumeral joints was markedly decreased and intraarticular pressure was raised during movements. The capsule showed hyperplasia of the synovial lining and vascular proliferation of the capsular wall. Functional and structural changes began to reverse after remobilization and returned to normal limits after 12 weeks.[27]

A more recent study, which looked at rat ankles immobilized for 2 to 6 weeks, found slightly different results. This study found that dense connective tissues remodel in such a way that mobility is unaffected after 2 weeks of im-

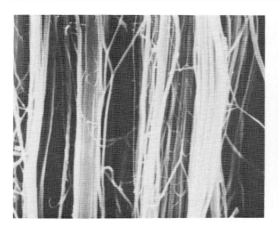 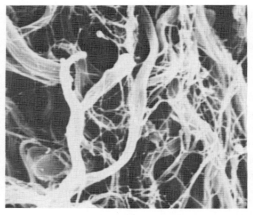

Figure 4–5 Electron micrograph of normal ligament (left) and healing scar at 2 weeks (right). *Source:* Reprinted from *Injury and Repair of the Musculoskeletal Soft Tissues* (p 112) by SL.-Y. Woo and J.A. Buckwalter with permission of the American Academy of Orthopaedic Surgeons, © 1987.

mobilization but markedly limited after 6 weeks of immobilization.[28] The authors attribute these changes to dense connective tissue undergoing remodeling between the 2 and 6 week periods. Earlier studies implied that cyclic mobilization of the immobilized joints caused rupture of the remodeled tissues, which limited early mobility. In Figure 4–6, following each yield point, the angle of the slope of the curve is unchanged. This supports the idea that rupture of the remodeled tissue that initially limited motion had not occurred; rather discrete adhesions between folds of tissues were responsible for this.

Langenskiold et al performed a study on immobilized, healthy rabbits. The authors found that casting for 5 to 6 weeks significantly decreased knee flexion. The resumption of normal activity, however, was able to restore 90% of joint mobility after 3 weeks. When immobilization was increased to 7 to 8 weeks, only 28% of knee flexion returned after 10 weeks of reconditioning. It took as long as 12 months for some of the animals to regain full mobility.[29] The study suggests that the longer the period of immobilization, the more difficult it becomes to regain normal tissue structure and mobility.

In a study performed by Evans et al,[22] experimentally immobilized rat knees were remobilized either by high-velocity manipulation, by range of motion, or both. The investigators found that, with manipulation, the macroadhesions were ruptured, and partial joint mobility was restored. If joint motion was allowed subsequent to the manipulation, functional range was regained.

Range of joint motion, along with freedom of movement, produced the same effect, although more gradually; after 35 days the joints were histologically indistinguishable. Rat knee joints immobilized for more than 30 days, however, did not regain full functional range. Again, the results suggest that movement restores the normal histological makeup of connective tissue, but the longer the period of immobilization, the lower the potential for achieving optimal results.

In summary, immobilization of connective tissue generally results in loss of ground substance with no net collagen loss (with immobilization periods of less than 12 weeks). The loss of ground substance also allows for significant water loss. Histologically, this results in decreased tissue extensibility due to the inability

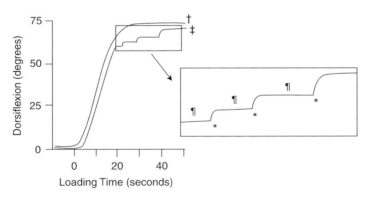

Figure 4–6 Diagrammatic representation of the qualitative difference in pattern of dorsiflexion between limbs casted for six weeks (‡) and all other limbs (†). In all ankles casted for 6 weeks, the curve exhibited intermediate plateaus (¶), followed by small but sudden slipping further into dorsiflexion (*), suggesting rupture of an adhesion with each slip. *Source:* Reprinted from Reynolds, C.A., Cummings, G.S., and Andrew, P.D. et al., The Effect of Nontraumatic Immobilization on Ankle Dorsiflexion, *Journal of Orthopaedic and Sports Therapy*, Vol. 23, No. 1, p. 31, with permission of the Orthopaedic and Sports Sections of the American Physical Therapy Association.

of the collagen fibers to maintain the critical interfiber distance, and the subsequent formation of microscopic collagen cross-links. At the macroscopic level, immobilization causes the formation of fibrofatty macroadhesions that become progressively more fibrotic with increased immobilization times. The studies also indicate that all periarticular connective tissues responded in the same basic fashion. Ligament and capsule surrounding fascia all had the same basic response to immobilization. Remobilization of the tissues causes a reversal of effects, provided the immobilization time has not been unreasonably long. More research is needed on duration of mobility and changes occurring within the connective tissues. Clinicians need to consider the early changes occurring in the immobilized connective tissues and adjust their treatment plans accordingly. Before 4 to 6 weeks, the stress-deprived, weakened cells may require gentle mid-range movement and protection from excessive forces; but after 6 weeks, treatment protocols should incorporate sufficient stress to induce connective remodeling to accommodate lengthened positions until full joint mobility is achieved.[28]

Traumatized Connective Tissue

Recently, questions have arisen about how traumatized connective tissue response to immobility differs from that of nontraumatized tissue. The previous studies have dealt with the response of nontraumatized connective tissue to immobilization. Some consider internal fixation of a limb to be a trauma-inducing form of immobilization, even though the fixation is located some distance from the tissue studied. In a study performed by Flowers and Pheasant,[30] human digits were casted for a period of several weeks and then examined. The range of motion lost during the immobilization period was regained within one treatment session of approximately 20 minutes. The implication of this study and of the previous immobilization studies is that when connective tissues of synovial joints are immobilized in the presence of inflammatory exudates, joint contractures occur, and result

from remodelling and shortening of connective tissues. When a limb is immobilized without inflammatory exudates being present, no contracture occurs, even after weeks.[5,6] Apparently, a catalyst is needed to begin the process of contracture—the catalyst is traumatic exudate. Also, methods of fixation may affect tissue changes.

The other factor in the different results reported in the two studies may be the method of fixation. The rigid fixation of the previous studies allowed for virtually no movement, whereas the cast fixation in the Flowers study may have allowed enough movement to prevent tissue changes. This phenomenon can be seen clinically for example, in the fixation methods of distal radial fractures. When the fracture is casted, a less than optimal union occurs, usually with the formation of extra callus. From a rehabilitation standpoint, the functional range of motion of the wrist, hand, and radio-ulnar joints is usually restored. If the fracture is fixated with an external fixator, the union is typically much cleaner, with less callus formation. Functional range of motion is typically not fully restored, however, especially in the wrist and radio-ulnar joints.

The clinical implications are threefold. First, patients entering physical therapy for rehabilitation following injury or surgery and subsequent immobilization will have connective tissue changes as just described. Second, a combination of two processes is occurring—scar formation and fibrosis. Scar formation occurs in areas that sustained direct insult and are in need of regeneration and repair. Fibrotic changes occur in tissues surrounding the scar area that were not directly traumatized but affected chemically by the traumatic exudates. Traumatic exudates infiltrate these surrounding, nontraumatized areas and, acting as chemical catalysts, create changes in the connective tissues.

Scar tissue versus fibrosis. Scar formation and fibrosis are two completely different histological processes, although some similarities exist. Scar formation is a localized response, with activity limited to a traumatized area, but fibrosis is a homogenous change in the "fabric" of the connective tissue. Limitation in mobility caused by scar tissue results from the lack of extensibility of the scar tissue and from the adhesions formed with adjoining healthy connective tissue. Limitation in mobility caused by fibrotic changes results from the lack of extensibility of the entire tissue. And as previously mentioned, fixation methods may play a part. Semirigid immobilization (immobilizer or cast) may allow sufficient movement to dampen the effects of immobilization.

For example, a shoulder may be frozen due to a macroscopic scar adhesion in the folds of the inferior capsule. A manipulation under anesthesia would tear the scar adhesion and restore mobility. A frozen shoulder may also be caused by a capsulitis, where the entire capsule shrinks (the analogy here is the size 5 capsule and a size 8 glenohumeral joint—the sock is simply too tight). The distinction is that homogenous changes in the capsule, rather than a single scar adhesion, limit motion. A manipulation under anesthesia may not be as successful in such a case, since an entire tissue is responsible for the immobility. The benefit of the increased mobility is outweighed by the potential damage to the capsule fabric and the restimulation of the fibrotic cycle.

Muscle Tissue

The response of muscle tissue to immobilization is less simplistic and more multifactorial than the response of connective tissue to immobilization. Being a contractile tissue, a muscle can be passively or actively immobilized and/or the muscle may be immobilized in a shortened or lengthened position. The muscle may be innervated or denervated, or predominantly slow twitch or predominantly fast twitch. Being a highly metabolic tissue, the immobilized muscle can undergo greatly varying metabolic changes, depending on its activity level. The purpose of this section is to outline briefly the histological response of muscle tissue to immobilization and to review the various factors influencing immobilized muscle that are the most applicable to myofascial manipulation.

One of the classic works on muscle response to immobilization was performed by Tabery et al.[31] In this study, cat soleus muscles were immobilized at various lengths and for various lengths of time. The animals were immobilized by plaster cast. Some of the animals were sacrificed and the muscles were biomechanically and histologically analyzed. Biomechanically, the passive length-tension was increased in the muscles immobilized in the shortened position, probably because of the connective tissue changes within and surrounding the muscle. Muscles immobilized in the lengthened position had no significant changes in the passive length-tension characteristics. From a histological standpoint, the muscles immobilized in the shortened position had a 40% loss of sarcomeres, with an overall decrease in fiber length. The muscles immobilized in the lengthened position exhibited a 19% increase in sarcomeres and an overall increase in fiber length. After 4 weeks of remobilization, the number of sarcomeres in the muscles returned to normal. This study illustrates the principle that muscle tissue will adapt to change in length by increasing or decreasing sarcomeres in order to keep sarcomeres at optimal lengths.

In a follow-up study performed by Tabery and Tardieu, muscle changes caused by prolonged *active* shortening were studied.[32] Sciatic nerves of guinea pigs were stimulated for 12 hours in either the shortened or lengthened position. The muscles stimulated in the shortened range had a 25% loss of sarcomeres after only 12 hours of contraction. Sarcomeres were completely recovered in the muscles between 48 and 72 hours. The implication of these studies is that muscles passively shortened lose sarcomeres at a much slower pace than muscles actively shortened.

Kauhanen et al immobilized the vastis intermedius of 13 rabbits in a shortened position for 2 to 28 days. After 3 days of immobilization, the muscle displayed a 15% decline in muscle fiber diameter. By 2 weeks, fatty changes were prominent and muscle fiber diameter had decreased to 56%. By 4 weeks, severe fibrotic damage of myofibrils was observed and fiber diameter had decreased to 47% of control values.[33]

Leivo et al[34] also immobilized the vastis intermedius of rabbits into the extended position. Progressive disorganization of myofibrils with breaking up of Z bands and an increase in the number and size of plasmic lipid vacuoles was seen with increased duration of immobilization. This study, as does the prior study, suggests that adverse mechanisms are in effect at the onset of disuse atrophy.

Kannus et al[35] found that, after 3 weeks of immobilization, there was a significant decrease in the mean percent of intramuscular connective tissue. They also found an increase in the relative number of muscle fibers with pathological alterations.

The clinical implication of these findings relates to the types of immobilization that occur in the practice setting. Immobilization may occur artificially (external or internal fixation), or as a physiological mechanism (muscle guarding). In the clinical setting, immobility may be due to trauma, past or present. A good example is the whiplash injury, in which immobilization is caused intrinsically by the cervical and upper thoracic paravertebral muscles, the scapulothoracic muscles, and the shoulder girdle muscles. In many cases, the surrounding musculature remains tonically active long after the facet or ligamentous strain has healed. The body learns a new recruitment pattern for the surrounding muscles, and this hypertonic pattern remains long after healing. The muscles are then actively "immobilized," causing some of the histological changes mentioned previously. Often, the most difficult part of the therapeutic process is dealing with this hypertonicity, which is secondary to the original injury.

PHYSIOLOGY OF MYOFASCIAL MANIPULATION

Massage has been used for centuries by various cultures around the world. Massage may be described as systematic, theraputic, and functional stroking and kneading of the soft tissues

of the body. The terms "myofascial manipulation" or "soft tissue mobilization" are used interchangeably with massage. In order to understand the effects of myofascial treatments on the body, a review of the available literature needs to be explored. Most studies on the effects of massage were published before the 1950s and were primarily animal studies. The effects discussed by these studies include circulatory changes, blood flow changes, capillary dilation, cutaneous temperature change, and metabolism. More recent studies, however, discuss the effects of massage on collagen and scar healing.

Effects of Massage on Blood Flow and Temperature

The effects on blood flow in the extremities of 17 adult men and women were analyzed by Wakim.[36] Groups were subdivided into those with no medical problems, those with rheumatoid arthritis, those with flaccid paralysis, and those with spasmatic paralysis. The subjects received two types of massage: (1) a *moderate depth* stroking and kneading massage described as a modified Hoffa-type massage, and (2) a *deeper* vigorous, stimulating, kneading, and percussion massage (as practiced in some European schools of physical therapy). The treated areas were the upper and lower extremities, and the massage lasted 15 minutes.

Wakim concluded that there was a consistent and significant increase in total blood flow and cutaneous temperature after deep stroking and kneading massage of the extremities in normal subjects, patients with rheumatoid arthritis, and subjects with spasmatic paralysis. A much milder effect was noted with the more superficial Hoffa-type massage and primarily in the group with paralysis. The greatest increase in circulation after deep stroking and kneading massage to the extremities occurred in subjects with flaccid paralysis. Significant increases in blood flow and temperature were still apparent in all groups receiving the deep massage when these signs were remeasured at 30 minutes. Blood-flow increases diminished markedly after

30 minutes. Neither deep stroking, kneading, nor vigorous stimulating massage of the extremities resulted in consistent, significant changes in blood flow of the contralateral unmassaged extremity.

The primary significance of Wakim's study is that temperature change in the extremities, resulting from increased blood flow to the part, may well depend on the manner in which the massage is administered. Wakim found that the moderate depth Hoffa massage affected only the blood flow of subjects with flaccid paralysis, whereas the deep stimulating massage had the effect of increasing the blood flow of all subjects studied.

Effects of deep kneading massage on venous blood flow were also examined by Wolfson, using animal models (dogs).[37] Massage was applied to the limbs above and below the knee (after anesthetizing), and was described as a "deep, kneading type of massage." Wolfson measured blood flow by cannulation of the femoral vein during anesthesia. The blood draining out was measured and reinjected into the opposite limb at the same rate the blood was being removed. The massage initially caused a fairly rapid increase in blood flow followed by a decrease in blood flow to a rate less than normal. This decrease in blood flow continued throughout the administration of the massage. Immediately following cessation of the massage, blood flow slowly returned to normal. Thus, Wolfson concluded that massage causes an increase in the rate of blood flow by mechanically emptying the blood vessels and allowing them to refill with fresh blood.

The findings in these studies are similar. Deep kneading massage increases the blood flow to the area being treated, as performed on human as well as animal models. Caution should be exercised, however, when comparing the results of animal studies to the human population.

The reaction of normal blood vessels to mechanical stimuli was microscopically examined by Carrier.[38] Gross visual observation of skin reaction was made following mechanical stimulation of the skin by a blunt instrument. With light stroking, the area in the path of the stroke blanched after a latent period of 15 to 20 seconds. The blanching lasted for several minutes. A harder stimulus resulted in a hyperemic line in the immediate path of the stimulus. With microscopic investigation, light pressure resulted in instantaneous opening of all capillaries in the microscopic field. A heavier pressure opened the underlying capillary for a longer unspecified duration.

Carrier's observations may correspond with the results of the studies by Wakim and Wolfson. If the moderate depth Hoffa-type massage (non-stimulating) is similar to the light stroke produced by a blunt instrument in Carrier's study, an immediate superficial capillary reaction is an effect of massage. The light stroke or Hoffa massage creates capillary dilation but for too short of a duration to affect blood volume, blood flow, or temperature in the underlying stroked area. When vigorous stimulating massage is administered, the result is a longer lasting dilation of the underlying capillaries, which creates a change in both blood flow and skin temperature. Both the vigorous stimulating massage and lighter Hoffa-type massage are used in myofascial manipulation.

In other research, Pemberton described the work of Clark and Swenson, who studied the capillary circulation in the ear of a rabbit following massage.[39] A permanent window was surgically created in the rabbit's ear, allowing observation of the capillaries. Following massage, an increase in rate of blood flow as well as actual changes in the vessel walls was noted. The vessel wall change was evidenced by the "sticking" and emigration of leukocytes. Clark and Swenson concluded that massage is accompanied or followed by an increased interchange of substances between the bloodstream and the tissue cells. The vessel wall change heightens the tissue metabolism.

Although massage is not defined in Clark and Swenson's study, the findings of increased blood flow and vessel wall change support the notion that massage, or soft tissue mobilization, affects the vascularization of the region underlying the massage. Clark and Swenson's conclusion agrees with Carrier,[38] who found an immediate capillary reaction underlying the stimulus of light and heavy pressure. Cutaneous temperature of an extremity following modified Hoffa massage was studied by Martin and associates.[40] They studied healthy adults and those with rheumatoid arthritis. Length of massage varied from 5 to 10 minutes.

Cutaneous temperature of the digits was measured with thermocouples. The results indicated that after massage of an extremity, there were superficial cutaneous temperature increases in the extremity lasting from 15 to 90 minutes. In a related investigation, the peripheral cutaneous temperature was examined after back massage. With three subjects, massage caused no change in the cutaneous temperature of the extremities.

Despite design and variable differences, the studies presented all agree on one point: massage causes capillaries to dilate in the region underlying the massage. If capillary dilation occurs, increased blood volume and flow occur, resulting in an increased temperature in the area of the massage.

Effect of Massage on Metabolism

Massage can also affect the metabolic processes, including the vital signs and bodily waste products. A review of the literature on massage's effect on human metabolism was performed by Cuthbertson.[41] Cuthbertson concluded that there was increased output of urine after massage, especially following abdominal massage. The excretion of acid was not consistently altered and there was no change in nitrogen content, inorganic phosphorus, or sodium chloride. The increased urine output occurred within 3 hours of the massage; the total net output of urine in a 24-hour period was unchanged. Among healthy patients in the survey, there was no increase in basal consumption of oxygen, pulse rate, or blood pressure. The above metabolic effects apply to a systemic process. Localized increase in basal consumption may occur, although localized effects have been inconclusive.

Because massage does not influence the basal metabolism, a likely explanation for the increased urine output is the massage's effect on the circulation of the part concerned. Increased blood volumes and blood flow through the area being massaged may cause the area to dispose of excess fluids during and after massage, thereby increasing urine output.

A recent study has also examined the benefits of massage on the human immunodeficiency virus (HIV) positive population. Twenty-nine gay men (20 HIV+, 9 HIV-) received daily massage for 1 month. After the 1 month of massage, a significant increase in the number of natural killer cells was noted in the HIV-positive men. Thus, there appears to be an enhancement of the immune system's cytotoxic capacity associated with massage. Further research in this area is required.[42]

The findings of increased blood flow, increased temperature, and increased metabolism to the area being massaged have strong clinical implications. They support the notion that massage is definitely indicated in areas where increased tissue circulation and nutrition are desired.

Physiological Reflexive (Autonomic) Effects of Massage

The literature on the reflexive, or autonomic, effects of massage consists of studies showing the effects of connective tissue massage distal to the area being treated. In support of connective tissue massage, Ebner reported that viscera, blood vessels, and supporting tissue and muscle cannot function as separate entities. Connective tissue massage stimulates the circulation to an area of the body that in turn, reflexively opens up increased circulatory pathways to other regions of the body. The cause for the initial increase in circulation is secondary to the mechanical tension created by the connective tissue massage strokes, which thereby stimulates the tissue.

Ebner studied the skin temperature of three patients after connective tissue massage.[43] Ebner found an increase in skin temperature (1°C to 2°C) of the foot following 20 minutes of connective tissue massage, which was performed on the sacral and lumbar segments of the back. Volker and Rostovksy (as reported by Ebner) also carried out experiments using connective tissue massage and found a maximum increase in temperature approximately 30 minutes after the massage ended distal to the area being massaged.

The mechanical friction of the massage stroke stimulates the structures within the connective tissue, primarily the mast cell. As the mast cell is stimulated, it produces histamine, which is a vasodilator. The vasodilation increases blood flow to the area treated and to other areas receiving histamine through the bloodstream. The increased permeability of the capillaries and small venules allows for quicker and more complete diffusion of waste products from the tissues to the blood. The blood components, when filtered by the kidney and excreted as urine, show increased nitrogen content, inorganic phosphorus, and sodium chloride, as reported by Cuthbertson.[41] The increased circulation caused by connective tissue massage (stimulating massage) through the reflexive nature of histamine release, follows the findings of Carrier, Martin et al, and Wakim when stimulating massage is

performed. Chapter 2 fully elaborates on the autonomic effects of myofascial manipulation.

Effects of Massage on Fibroblastic Activity/ Collagen Synthesis during the Healing Process

Research has shown that controlled motion of soft tissues influences the healing process.[44–47] As discussed prior, the soft tissues of the body are subjected to both internally and externally generated forces. Without stress applied through the tissues, the tensile strength will decrease.[47] Stearns[48] observed the effect of movement on the fibroblastic activity in the healing connective tissues. She concluded that fibrils form almost immediately. External factors were responsible for assuming an orderly arrangement of these fibrils. Cyriax and Russell[49] believe that gentle passive movements of the soft tissues will prevent abnormal adherence of the fibrils without affecting their proper healing.

The manual therapist should use his or her knowledge of the stages of healing to determine when specific massage techniques should be utilized (Figure 4–7). The previous chapter dis-

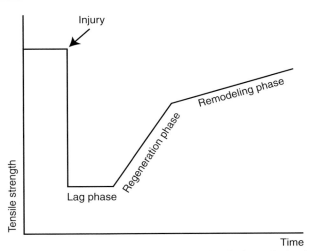

Figure 4–7 General trend of increase in tensile strength of injured soft tissue during healing process. *Source:* Reprinted with permission from P. Evans, The Healing Process at the Cellular Level, *Physiotherapy*, Vol. 66, No. 8, pp. 256–259, © 1980, Physiotherapy Canada, and G. Hunter, Specific Soft Tissue Mobilization in the Treatment of Soft Tissue Lesions, *Physiotherapy*, Vol. 80, No. 1, pp. 15–21, © 1994, Physiotherapy Canada.

cussed the soft tissue's inability to withstand stress immediately after injury. It is, therefore, important to protect the injured tissues from stress during the early inflammatory stage. The fibrin bond holding the wound together can easily be disrupted, ultimately leading into an increase in the amount of scar tissue formed.[50] As collagen does not appear in the wound for 4 to 6 days after injury, the value of friction or deep massage before this time is questionable.[51]

As the tissues move into the regeneration phase, fibroblasts begin to lay down collagen, and the tensile strength increases. Recent research using augmented soft tissue mobilization (ASTM) has proven to be effective during this stage. ASTM uses specially designed instruments to assist the therapist in mobilization of soft tissue fibrosis. An animal model using rat Achilles tendon injuries revealed that ASTM leads to an increase in fibroblast recruitment and activation as well as an increase in fibronectin production.[52] By increasing fibroblast activity, the healing process in this animal model was enhanced.[53] Carefully applying tension during this phase will help collagen fibers to align properly.[54,55] Transverse friction at this point can be gently begun as not to detach the healing fibers. The transverse movement is an imitation of the muscle's normal mobility by broadening but not stretching or tearing the healing fibers.[56] The movement will encourage realignment and lengthening of fibers.

As the remodeling phase begins, collagen synthesis equals collagen lysis. Evans[50] found that collagen fibers tend to contract and decrease scar tissue mobility at this point. Collage cross-linking adds strength to the wound but can also lead to a decrease in mobility. During this phase, the wound should be continually tensioned to promote good fiber orientation and scar tissue extensibility. The use of deep massage techniques may be appropriate at this time to decrease adhesions and break down scar.[49] One study on friction massage done for 10 minutes a day over 3 months on pediatric burn patients with hypertrophic scarring failed to show any increase in pliability or height of the scar.[57] Further studies using longer or more frequent treatment sessions should be done before concluding that massage is ineffective in the treatment of hypertrophic scarring.

CONCLUSION

The literature supports the use of myofascial techniques to influence the healing of soft tissues. The choice of technique by the physical therapist should be based in part on the stage of healing of the injured tissue. Gentle techniques may be beneficial early on to ensure an orderly arrangement of fibrils and to prevent adhesions. In the latter stages of healing, deeper techniques may be more appropriate in order to decrease adhesions, improve scar extensibility, and increase overall mobility of the soft tissues. A good manual therapist must not only understand the histopathology of myofascia and the stages of healing but must remember to use this knowledge when choosing treatment techniques. Choosing the appropriate technique at the appropriate time is essential to successful treatment.

REFERENCES

1. Hardy M. The biology of scar formation. *Phys Therapy*. 1989;69(12):22–30.

2. Norris C. *Sports Injuries: Diagnosis and Treatment for the Physical Therapist*. Oxford: Butterworth-Heinmann Ltd; 1993:21–24.

3. VanDer Muelen JCH. Present state of knowledge on processes of healing in collagen structures. *Int J Sports Med*. 1982;3:4–8.

4. Kellett J. Acute soft tissue injuries—A review of the literature. *Med Sci Sports Exer*. 1986;18(5):489–500.

5. Cummings GS. *Soft Tissue Contractures: Clinical Management.* Continuing Education Seminar. March 1989. Course Notes. Georgia State University.

6. Cummings GS, Crutchfield CA, Barnes MR. *Orthopedic Physical Therapy Series: Soft Tissue Changes in Contractures.* Atlanta, GA: Stokesville Publishing; 1983.

7. Lachman S. *Soft Tissue Injuries in Sports Medicine.* Oxford: Blackwell Publishing; 1988.

8. Hettinga DL. Inflammatory response to synovial joint stiffness. In: *Orthopedic and Sports Physical Therapy,* 2nd ed. St Louis: CV Mosby; 1990.

9. Leibovich SJ, Ross R. The role of macrophages in wound repair. *Am J Pathol.* 1975;78:71–79.

10. Lotz M, Duncan M, Gerber L. Early versus delayed shoulder motion following axillary dissection. *Ann Surg.* 1981;193:288–295.

11. Paletta F, Shehadi S, Mudd J. Hypothermia and tourniquet ischemia. *Plas Reconstruct Surg.* 1962;29:531–538.

12. Arem A, Madden J. Effects of stress on healing wounds. *J. Surg Res.* 1976;20:93–102.

13. Kellett J. Acute soft tissue injuries—a review of the literature. *Med Sci Sports Exer.* 1986;18(5):489–500.

14. Videman T, Eronen I, Friman C, et al. Glycoaminoglycan metabolism of the medial meniscus. *Acta Orthop Scand.* 1979;50:465–470.

15. Videman T, Michelsson J, Rauhamaki R, Langenskiold A. Changes in S-sulfate uptake in different tissue in the knee and hip. *Acta Orthop Scand.* 1976;47:290–298.

16. Woo S, Matthews JV, et al. Connective tissue response to immobility. *Arthritis Rheum.* 1975;18:257–264.

17. Akeson WH, Woo SL-Y, et al. The connective tissue response to immobilization: biochemical changes in periarticular connective tissue of the rabbit knee. *Clin Orthop.* 1973;93:356–362.

18. Akeson WH, Amiel D, et al. The connective tissue response to immobility: an accelerated aging response. *Exp Gerontol.* 1968;3:289–301.

19. Akeson WH, Amiel D, et al. Collagen cross-linking alterations in the joint contractures: Changes in the reducible cross-links in periarticular connective tissue after nine weeks of immobilization. *Connect Tissue Res.* 1977;5:l5–l9.

20. Akeson WH, Amiel D. Immobility effects of synovial joints: The pathomechanics of joint contracture. *Biorheology.* 1980;17:95–110.

21. Akeson WR, Amiel D. The connective tissue response to immobility: A study of the chondroitin 4 and 6 sulfate and dermatan sulfate changes in periarticular connective tissue of control and immobilized knees of dogs. *Clin Orthop.* 1967;5l:190–197.

22. Evans E, Eggers G, et al. Experimental immobilization and mobilization of rat knee joints. *J Bone Joint Surg.* 1960;42A:737–758.

23. Neuberger A, Slack H. The metabolism of collagen from liver, bones, skin, and tendon in normal rat. *Biochem J.* 1953;53:47–52.

24. Schiller S, Matthews M, et al. The metabolism of mucopolysaccharides in animals: Further studies on skin utilizing C14 glucose, C14 acetate, and S35 sodium sulfate. *J Biol Chem.* 1956;218:139–145.

25. Schiller S, Matthews M, et al. The metabolism of mucopolysaccharides in animals: Studies in skin using labeled acetate. *J Biol Chem.* 1955;212:531–535.

26. Amiel D, Akeson WH, et al. Stress deprivation effect on metabolic turnover of medial collateral ligament collagen. *Clin Orthop.* 1983;172:265–270.

27. Schollmeier G, Sarkar K, Fukuhara K, et al. Structural and functional changes in the canine shoulder after cessation of immobilization. *Clin Orthop.* 1996;323:310–315.

28. Reynolds CA, Cummings GS, Andrew PD, et al. The effect of nontraumatic immobilization on ankle dorsiflexion. *JOSPT.* 1996;23(13):27–33.

29. Langenskiold A, Michalsson JE, Videman T. Osteoarthritis of the knee in the rabbit produced by immobilization. *Acta Ortop Scand.* 1979;50:1–14.

30. Flowers KR, Pheasant SD. The use of torque angle curves in the assessment of digital stiffness. *J Hand Therapy.* 1988;January–March:69–74.

31. Tabery JC, Tabery C, et al. Physiological and structural changes in the cat's soleus muscle due to immobilization at different lengths by plaster casts. *Am J Physiol.* 1972;224:231–244.

32. Tabery JC, Tardieu C. Experimental rapid sarcomere loss with concomitant hypoextensibility. *Muscle Nerve.* 1981;May/June:198–203.

33. Kauhanen S, Leivo I, Petilla M, et al. Recovery of skeletal muscle after immobilization of rabbit hind limb. *APMIS.* 1986;104(11):797–804.

34. Leivo I, Kauhanen S, Michaelsson JE. Abnormal mitochondria and sarcoplasmic changes in rabbit skeletal muscle induced by immobilization. *APMIS.* 1998;106(12):1113–1123.

35. Kannus P, Jozsa L, et al. Free mobilization and low-to high-intensity exercise in immobilization-induced muscle atrophy. *J Applied Physiol.* 1998;84(4):1418–1424.

36. Wakim KG. The effects of massage on the circulation in normal and paralyzed extremities. *Arch Phys Med Rehabil.* 1949;30:135.

37. Wolfson H. Studies on effect of physical therapeutic procedures on function and structures. *JAMA.* 1931;96:2020.

38. Carrier EB. Studies on physiology of capillaries: Reac-

tion of human skin capillaries to drugs and other stimuli. *Am J Physiol.* 1922;61:528–547.

39. Pemberton R. Physiology of massage. In: *AMA Handbook of Physical Medicine and Rehabilitation.* Philadelphia: Blakinston Co; 1950:133.

40. Martin GM, Roth GM, et al. Cutaneous temperature of the extremities of normal subjects and patients with rheumatoid arthritis. *Arch Phys Med Rehabil.* 1946; 27:665.

41. Cuthbertson DP. Effect of massage on metabolism: A survey. *Glasgow Med J.* l933;2:200–2l3.

42. Ironson G, Field T, Scafidi F, et al. Massage therapy is associated with enhancement of the immune systems cytotoxic capacity. *Int J Neurosci.* 1976;84(1–4):205–217.

43. Ebner M. *Connective Tissue Manipulations.* Malabar, FL: Robert E Kreiger Publishing; 1985.

44. Takai S, Woo SLY, Horibe S, et al. The effects of frequency and duration of mobilization on tendon healing. *Journal Orthop Res.* 1991;9(5):705–713.

45. Frank C, Akeson WH, Woo SLY, et al. Physiology and therapeutic value of passive joint range of motion. *Clin Orthop Rel Res.* 1984;185(5):113–125.

46. Gomez MA, Woo SLY, Amiel D, et al. The effects of increased tension on healing medial collateral ligaments. *Am J Sports Med.* 1991;19(4):347–354.

47. Forrester J, Zederfeldt B, Hayes T, et al. Wolff's law in relation to healing skin. *J Trauma.* 1970;10(9):770–779.

48. Stearns ML. Studies of the development of connective tissue in transparent chambers in the rabbit ear. *Am J Anat.* 1940;67:55–97.

49. Cyriax J, Russell G. *Textbook of Orthopedic Medicine,* volume 2. London: Tindall and Cassall Ltd; 1990.

50. Evans P. The healing process at the cellular level. *Physiotherapy.* 1980; 66(8):256–259.

51. Hunter G. Specific soft tissue mobilization in the treatment of soft tissue lesions. *Physiotherapy.* 1994;80(1): 15–21.

52. Davidson CJ, Ganion LR, Gehlsen G, et al. Rat tendon morphological and functional changes resulting from soft tissue mobilization. *Med Sci Sports Exercise.* 1997; 29(3):313–319.

53. Melham TJ, Sevier TL, Malnofski MJ, et al. Chronic ankle pain and fibrosis successfully treated with a new non-invasive augmented soft tissue mobilization technique. *Med Sci Sports Exer.* 1998;40(6):801–804.

54. Tipton CM, Mathes RD, Maynard JA, et al. Influence of physical activity on ligaments and tendons. *Med Sci Sports.* 1975;7:165–175.

55. Postacchini F, Demartino C. Regeneration of rabbit calcanial tendon maturation of collagen and elastic fibers following partial tenotamy. *Connect Tissue Res.* 1980; 8:41–47.

56. Chamberlain G. Cyriax friction massage. *JOSPT.* 1982; 4(1):16–22.

57. Patino O, Novick C, Merlo A, et al. Massage in hypertrophic scar. *J Burn Care Rehabil.* 1999;20(3):268–271.

CHAPTER 5

Neuromechanical Aspects of Myofascial Pathology and Manipulation

Clayton D. Gable

The mere motion of muscular and/or fascial tissues through stretching feels good to humans and many other vertebrate animals. One has only to think about their own tendency to stretch on awakening in the morning or after a long trip by airplane or car. Even animals such as our pets seem to like stretching. Walsh cited E. K. Borthwick, Emeritus Professor of Classics at Edinburgh University, for the following account:

The verb "stretch" (τεινω, teino) is the common form and is used by Homer of stretching of a bow, reins, etc.—"to stretch oneself in running." Aeschylus uses it of straining the voice. Galen uses it of stretching tendons, etc.

The noun, τονοσ (tonos), is apparently attested in Xenophanes (sixth century BC philosophic poet) of exertion or striving after virtue or courage. It is used by Aeschylus of stretching flax; in Herodotus and Aristophanes of bed and chair cords, in Plato and Aeschines, of pitch of voice, or accent; in Aristoxenus and subsequent musical writers of pitch-key; in the medical writer Soranus (second century AD) of power of contracting muscles.[1(p6)]

As one can surmise from the passage above, muscle contraction has, for almost 2000 years, been associated with stretching.

Given that stretching is such an integral part of normal human and vertebrate behavior and the 100-year history of study of the influence of various sensory mechanisms on movement, it is necessary to review some neurology that is associated with myofascial tissues. To that end, this chapter reviews the basic neurology of myofascial tissues emphasizing the afferent or stimulus perception side of the equation. In addition, the author reviews some of the more contemporary findings regarding (1) the influence of somatosensory receptors on movement control, (2) muscle "tone," and (3) the interaction of biomechanical properties of myofascial tissues and the nervous system.

Following the review of the basic science regarding neurology and movement control, there is a science/application section. *This section offers explanations for some of the techniques found in Part III of this volume in terms of current understanding of the reviewed neuroscience and neuromechanical aspects of myofascial tissues.*

BASIC AFFERENT NEUROLOGY OF CONNECTIVE TISSUE

A detailed presentation of the state of current neuroscience of receptor anatomy and physiology is beyond the scope of this book. Therefore, the following information summarizes classical and recent understandings of peripheral receptors in skin and the various connective tissues of myofascia. These receptors fall into four major categories of mechanoreceptors, nociceptors, thermoreceptors, and chemoreceptors. All of these receptors influence or are influenced by movement, temperature, physiology, or pathology. Also, all of these receptors have influence on movement and movement control as well as direct and indirect influences on cardiovascular and respiratory physiology.

Mechanoreceptors

Mechanoreceptors are exactly what the name implies; they are peripheral sensory receptors of mechanical events. They transduce mechanical energy into nerve impulses, which are then transmitted to the central nervous system via their afferent neuron axons. They are located throughout the musculoskeletal system, the vascular tree, and the skin. They include specialized neuronal structures and free nerve endings (Table 5–1).

Each of the various mechanoreceptors listed in Table 5–1 has particular anatomies, firing characteristics, thresholds, conduction velocities and, most importantly of all for a clinician, *functional and physiologic effects.* Therefore, the next few sections review some of the pertinent characteristics and functional implications of

Table 5–1 Mechanoreceptors

Receptor Type	Fiber Size and Group	Location and Information Transduced
Meissner's corpuscle	Aβ	Skin: touch
Pacinian corpuscle	Aβ	Skin: flutter Fibrous connective tissue: compressive stimuli
Ruffini's corpuscle	Aβ	Skin: steady indentation Fibrous connective tissue: tension on structures such as ligaments
Merkel's receptor	Aβ	Skin: steady indentation
Hair-guard, hair-tylotrich	Aβ	Skin: steady indentation
Hair-down	Aβ	Flutter
Primary muscle spindle	Aα Ia	Dynamic change of length
Secondary muscle spindle	Aβ II	Muscle length, mostly static
Golgi tendon organ	Aα II	Tension on a tendon
Joint capsule receptors (Type II)	Aβ II	Extremes of joint position (i.e., maximum tension on joint capsule)
Muscle afferents (III)	Aδ III	Mechanical, chemical, and thermal stimuli in muscle
Muscle afferents (IV)	C IV	Mechanical, chemical, and thermal stimuli in muscle
Bare nerve endings	A–C	Mechanical chemical, thermal, and pain

the various receptors. It is important to note that even those receptors listed in Table 5–1 as being primarily located in the skin contribute to proprioception and kinesthesia. Gardner, Martin, and Jessell state the following.

> Three types of mechanoreceptors in muscle and joints signal the stationary position of the limb and the speed and direction of limb movement: (1) specialized stretch receptors in muscle termed muscle spindle receptors; (2) Golgi tendon organs, receptors in the tendon that sense contractile force or effort exerted by a group of muscle fibers; and (3) receptors located in joint capsules that sense flexion or extension of the joint."[2(p443)]

It is with Gardner, Martin, and Jessell's statement in mind that the following review is offered.

Meissner's Corpuscles

Meissner's corpuscles are specialized structures located in glabrous (hairless) skin (e.g., palms, soles of feet, lips) of mammals. They are rapidly adapting in their response to mechanical stimuli such as skin indentation (Figure 5–1.) The *rapidly adapting* characteristic is common to several skin mechanoreceptors. It indicates that a rapidly adapting receptor will respond to a stimulus event with *an* action potential, and then the receptor will go silent for a period of up to several seconds failing receipt of another mechanical event.

In the case of a Meissner's corpuscle, a single indentation of 70 to 1000 micrometers (0.00007–0.01 mm) into the skin would result in a single action potential with a subsequent silent period of up to several seconds. Although this behavior would appear to be somewhat dysfunctional, rapidly adapting receptors have another characteristic. They are responsive to repetitive

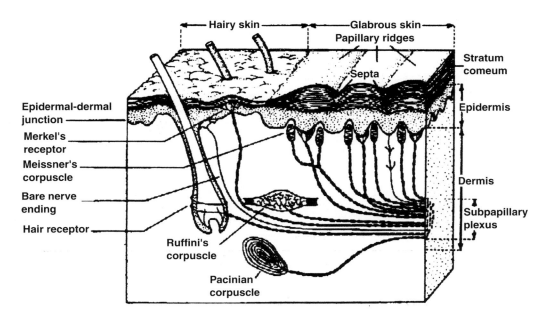

Figure 5–1 Receptors in hairy and hairless skin. *Source:* Reprinted with permission from J.H. Kandel et al., eds., *Principles of Neural Science*, 3rd ed., pp. 533–547, © 1991, McGraw-Hill Companies.

stimuli (at varying frequencies) with repeated action potentials.

Meissner's corpuscles, specifically, respond to repetitive stimuli, such as sinusoidal indentations of the skin, at frequency ranges of 2 or 3 Hz up to around 300 Hz. Compared with Pacinian corpuscles, this range is a relatively slow frequency range. As previously mentioned, this range of stimulus indentation is from 70 to 1000 micrometers, with the greatest sensitivity at between 10 and 100 Hz of stimulus (Figure 5–2). With a rapidly adapting system, the perception of relatively low frequency and low amplitude indentations of the skin is possible. In particular, the density of Meissner's corpuscles is higher in glabrous skin of such structures as the hands.[2] This is most beneficial for the therapist in palpation and during treatment. The property of rapid adaptation gives the Meissner's corpuscles excellent temporal resolution in perception of rapid and subtle change. It does nothing, however, to explain their superior spatial sensitivity.

There are two other characteristics for which their superior spatial resolution may be accountable. First, Meissner's corpuscles are mechanically coupled to the surrounding subcutaneous tissues by thin strands of connective tissue. These strands promote the transmission of adequate stimulating force to several surrounding corpuscles for a given pinpoint stimulus area. The second characteristic is related partially to this mechanical coupling but mostly to the fact that the *receptive field* for Meissner's corpuscles is very small (2–4 mm in diameter).

A *receptive field* can be thought of as an isolated area of skin that can be stimulated and the area that is perceived to be stimulated. In an area of skin with small receptive fields, stimulus of a small point results in perception of stimulus that is restricted to just that small point. Conversely, an area with large receptive fields will result in perception of stimulus to a large area, even with only a small point stimulated.

The impact of Meissner's corpuscles on practitioners of manual therapeutic technique would be difficult to overstate. With their excellent spatial resolution and ability to perceive relatively small differences in texture, tissue density, and so forth, the manual practitioner certainly utilizes them in all of his or her practice. Other

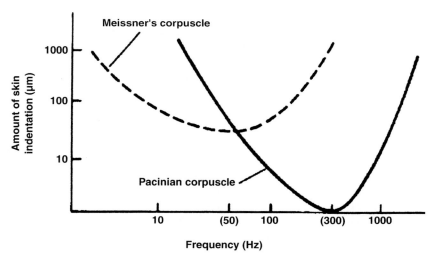

Figure 5–2 Sensitivity to skin indentation. *Source:* Reprinted with permission from J.H. Kandel et al., eds., *Principles of Neural Science*, 3rd ed., pp. 533–547, © 1991, McGraw-Hill Companies.

implications for the Meissner's corpuscles are discussed in a later section of this chapter concerning effects and interactions of connective tissue neurophysiology with movement control.

Pacinian Corpuscles

Pacinian corpuscles are located in the subcutaneous tissue of both hairy and glabrous skin. Although the skin is probably the largest organ with the greatest density of Pacinian corpuscles, it certainly is not the only location. As early as 1882, Hagen-Torn found Pacinian corpuscles in the joint capsules of various animals.[3] Gardner investigated joint capsules further and with better technology. He failed to find Pacinian corpuscles in the articular capsule but did find them in the fibrous periosteum near articular or ligamentous attachments.[4] Zimny et al reported finding Pacinian corpuscles, in the anterior and posterior horns of knee menisci, accompanied by Ruffini endings, free nerve endings, and Golgi tendon organs.[5]

A great deal is known about Pacinian corpuscles concerning their anatomy and function. They consist of a specialized nerve ending that is surrounded by connective tissue laminae. This connective tissue laminae makes the corpuscle a rapidly adapting receptor, which makes it responsive to stimuli at frequencies from 15 to 1000 Hz. This rapidly adapting quality allows the corpuscles to be sensitive to rapid change in stimulus intensity. As with other mechanoreceptors, the Pacinian corpuscle is sensitive to mechanical energy. In fact, it is extraordinarily sensitive (down to a level of less than 1-micrometer skin indentation, which will result in an action potential from the Pacinian corpuscle). Even though Pacinian corpuscles are very sensitive to mechanical energy, their very large receptive fields make them exceedingly poor for localization. Recall that the receptive field of a Meissner's corpuscle is from 2 to 4 mm in diameter with excellent localization. In contrast, the receptive fields of Pacinian corpuscles are so large that in experimental stimulation of Pacinian corpuscles, humans were able to localize only to one finger or to the medial half of the palm.[2]

Ruffini Corpuscles

Ruffini corpuscles are found in the subcutaneous tissue beneath both hairy and glabrous skin.[2] They are also found in the superficial layers of fibrous joint capsules and other connective tissue surrounding joints.[6] Their intertwining with the connective tissue is functional in that they are stimulated by the displacement of the collagen fibers surrounding them. They are slowly adapting receptors and they also have very large receptive fields. One major advantage of their slowly adapting characteristic is of functional significance. Since they do not "turn off" following a stimulus but continue to fire with a consistently applied stimulus, they contribute to steady-state position sense and tactile sensation.

Hair Receptors

Hair receptors are divided functionally into two categories based principally on the type of stimulus to which they respond. Tylotrich (stiff) hair receptors are responsive to steady skin indentation, and down hair receptors are sensitive to flutter. Basically, hair receptors are specialized nerve endings incorporated in the connective tissue at the base of a follicle and are very sensitive to mechanical deformation of the hair. Their implications for clinical practice of the manual therapist are most likely restricted to an awareness of their presence and the knowledge that they, as with most any receptor, can be sensitized under conditions of paraesthesia.[7]

Merkel's Receptors

Merkel's receptors are probably the most peripheral of all the sensory receptors. They are located in the epidermis of glabrous (hairless) skin. They have unusual receptors in that the receptors appear to synapse with epithelial cells. This synapse or connection of epithelial cells directly with the Merkel's receptors results in an action potential for the neuron serving the receptor with any mechanical stimulus to its related epithelial cell(s). Merkel's receptors are slowly adapting like Ruffini endings, but unlike them, Merkel's receptors have very small receptive fields. Therefore, these endings can respond to very small stimuli and localize well with their 2- to 4-mm sized fields along with their capacity to continue to send "tonic" signals to the central nervous system without a change in stimulus intensity.[2]

Muscle Spindles

Muscle spindles are located in striated (skeletal) muscle and transduce changes in length of the muscle. They fall into three categories and are named for their shape and function. The *dynamic nuclear bag fibers* transduce information about rapid changes in length, the rate of change of length, and are most heavily concentrated in phasic muscles. The *static nuclear bag spindles* (so named because of the central bunching of the muscle fiber nuclei) transduce more tonic information about spindle length. *Nuclear chain fibers* (muscle fiber nuclei are evenly distributed throughout the muscle fiber) transduce information about slower changes in length and are more concentrated in tonic muscles. The physiology of muscle spindles, as well as their influence on muscle tone, is worthy of repeat here (Figure 5–3).

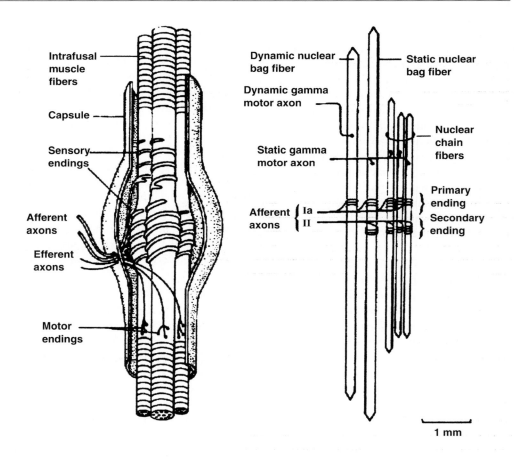

Figure 5–3 Components of muscle spindles. *Source:* Reprinted from *Trends in Neurosciences*, Vol. 3, I.A. Boyd, The Isolated Mammalian Muscle Spindle…, pp. 258–265, © 1980, with permission from Elsevier Science.

Muscle spindles are specialized encapsulated structures that are arranged in parallel with skeletal muscle fibers. They consist of (1) a group of special muscle fibers (*intrafusal fibers*), which are located in the spindle; (2) sensory axons that terminate as a spiral ending around the intrafusal fibers; and (3) motor axons (*gamma efferent fibers*), which adjust the sensitivity of muscle spindles. In order to understand the differential physiology of muscle spindles, it is necessary to discuss the two sensory endings (i.e., afferent fibers) that transduce length information from the spindles.

Primary Endings. Primary endings consist of branches of a Group Ia afferent axon. Primary endings terminate on all three types of intrafusal fibers in the muscle spindle. The spiral endings wrap around the central or equatorial region of the muscle spindles. The structure of the membrane of both primary and secondary endings contains stretch sensitive channels which, when activated, depolarize the axolemma of the primary and/or secondary endings sending an impulse up the afferents to the spinal cord. Primary endings exhibit a property known as *velocity sensitivity* where they increase their firing rate with a sudden and rapid change in length. This velocity sensitivity is manifested both with rapid shortening and with lengthening. In the case of rapid shortening, the primary endings lapse in their firing and then continue firing at a lower frequency when the shortening halts. With lengthening, the primary endings will demonstrate increased firing rates that are dependent upon the rate of the lengthening (stretching). Consequently, primary endings respond vigorously with increased firing rates with brief stimuli such as a tap or vibration. In addition to their sensitivity to velocity of stretch, they are also very sensitive to changes in length, demonstrating increased firing rates with stretches of as little as 0.1 mm.

Secondary Endings. Group II afferents branch into secondary endings, which innervate the static nuclear bag fibers and the nuclear chain fibers in a spiral fashion, like the primary endings. The secondary endings function similarly to the slowly adapting receptors already discussed. That is to say, secondary endings exhibit a fairly steady firing rate in either the presence or absence of movement. These slowly adapting receptors are, therefore, ideal for signaling the length of muscle without the need for movement to increase their sensitivity.

Gain Adjustment of Muscle Spindles. As one can surmise from the previous discussion and Figure 5–4, when the extrafusal muscle fibers of a muscle contract in response to a stimulus from an alpha motor neuron, the intrafusal fibers would be left slack. Therefore, there would be a silent period in the firing of both the primary and secondary endings. To eliminate this silent period, the *gamma motor neuron* system activates the intrafusal fibers and maintains the sensitivity of the spindles. When the gamma motor neuron system activates the intrafusal fibers simultaneously with the alpha motor neuron system, activating the extrafusal fibers, the sensitivity of the primary and secondary endings is maintained.[8]

Golgi Tendon Organs

Golgi tendon organs (GTOs) transduce information about tension and are located at the musculotendinous junctional zones. The GTOs transduce mechanical information through specialized nodes on free nerve endings. GTOs are free nerve endings with specialized nodes on their branches that respond to the mechanical deformation from collagen fibers placed on stretch. These free nerve endings are made up of branches of Ib afferent neurons that become unmyelinated after entry into the GTO. The structure of the collagen strands inside a GTO is in the form of a braided arrangement. In the same way that longitudinal tension approximates the fibers of a Chinese finger trap, tension applied to the muscle approximates the braided strands of the GTO and compresses the branches of the Ib afferent neurons (Figure 5–5). Golgi tendon organs are extremely sensitive to changes in ten-

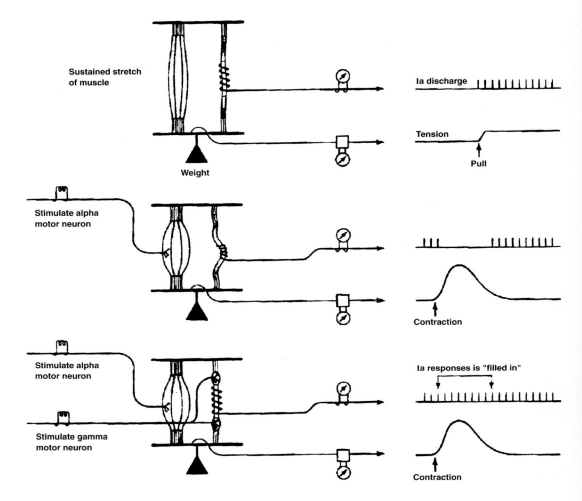

Figure 5–4 During active muscle contractions the ability of the spindles to sense length changes is maintained by activation of gamma motor neurons. (Adapted from Hunt and Kuffler, 1951.) (**A**) Sustained tension elicits steady firing of the Ia afferent. (**B**) A characteristic pause occurs in ongoing discharge when the muscle is caused to contract by stimulation of its alpha motor neuron alone. The Ia fiber stops firing because the spindle is unloaded by the contraction. (**C**) If during a comparable contraction a gamma motor neuron to the spindle is also stimulated, the spindle is not unloaded during the contraction and the pause in Ia discharge is "filled in." *Source:* Adapted with permission from C.C. Hunt and S.W. Kuffler, Stretch Receptor Discharges During Muscle Contraction, *Journal of Physiology*, Vol. 113, pp. 298–315, © 1951, The Physiological Society.

sion on the connective tissue in which they are located. This sensitivity has been documented at levels as low as the force generated by a twitch contraction of a single motor unit in the triceps surae of a cat (i.e., very few grams of force).[9]

Another important feature of the GTO is in their combination with muscle spindles. The reader will recall that the primary endings from dynamic nuclear bag fibers experience a pause in their firing during contraction of a muscle.

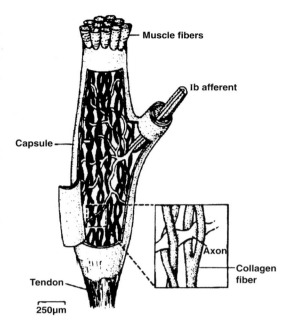

Figure 5–5 Golgi tendon organs. *Source:* Reprinted with permission from J.E. Swett and T.W. Schoultz, Mechanical Transduction in the Golgi Tendon Organ: A Hypothesis, *Archives de Italiennes de Biologie* Vol. 113, pp. 374–382, © 1975, Archives de Italiennes de Biologie.

Unlike them, the GTOs are highly active with contraction of a muscle secondary to the tension exerted on them by the muscle.

Implications of Muscle Spindles and Golgi Tendon Organs

While the impact of the alpha and gamma motor neuron system is quite well understood conceptually, if not in detail, by most practitioners of manual therapeutics, the impact of pathology in connective tissue may require some discussion. Intrafusal fibers are arranged in parallel to the extrafusal muscle fibers. Muscle spindles measure approximately 4 to 5 mm in length and 1 mm in diameter. With their parallel arrangement, they are connected to either end of their muscle's attachment by long collagen-containing fibers. Golgi tendon organs are arranged in series with the extrafusal muscle fibers

and are approximately 1 mm in length and 300 to 500 micrometers in diameter. Hence, a change in the mechanics of a muscle secondary to injury can change the firing patterns of either or both of these proprioceptors.

Consider the following scenario. An athlete sustains a contusion to the distal third of the medial head of the gastrocnemius muscle. This occurred 5 days ago with fairly good resolution of the edema. During gait, the gastrocnemius is active from mid-stance (as a decelerator) until toe-off (as an accelerator). Assuming a relatively normal foot posture and equal forces, rate of change of length, and length changes being generated by the medial and lateral head of the gastrocnemius, the afferent stimuli coming from the medial and lateral head of the gastrocnemius would, under nonpathologic conditions, be approximately equal. Under the current conditions of a contusion that is in the process of healing but having formed some scar tissue, however, the afferent information is different between the 2 heads of the gastrocnemius. In the medial head, the afferent information is altered because of scarring of the collagenous connections of the intrafusal fibers. This results in a mechanical "mis-link" from collagenous cross-bridges and scarring and results in a perceived change in length that is reflective of the actual change in length. In addition to the mismatch between the two heads of the gastrocnemius relative to the length of the muscles, there is a problem with tension information from the Golgi tendon organs.

The serial arrangement of the Golgi tendon organs makes them sensitive to tension generated along the mechanical chain of the muscle. Therefore, changes in the viscoelastic properties of the muscle to which it is attached can produce a differential in tension (particularly at the initiation of contraction). This differential tension produces another mismatch between the tendinous origin of the medial and lateral heads and even the possibility of differences within the fascicles of the medial head attaching to the Achilles tendon. With the decreased elasticity of the muscle from the collagenous cross-bridges and scarring, there is a relatively higher tension perceived by the medial head compared with the lateral head. In addition to all of the collagenous cross-bridges, which have affected the elasticity of the muscle, there is also the problem of changes in viscosity of the muscle (intrafusal and extrafusal fibers) and all the surrounding connective tissue (see Chapter 3 for a review of viscoelasticity). These changes in viscosity and elasticity contribute to an afferent information mismatch, when compared with the conditions under which most tasks are learned.

Joint Capsule Receptors

Joint receptors include quite an array of receptor types. Among them, the Ruffini endings are classified as Type I receptors by Wyke.[10] They tend to be more heavily concentrated in the proximal extremity joint capsules and are slowly adapting. This makes them ideally suited to provide postural information about static positions and their necessarily relatively stable positions to allow distal movement.

Pacinian corpuscles are Type II receptors and are located in the deeper layers of the joint capsule and the fat pad. They also tend to be concentrated near the bony attachments of the joint capsule. Of interest is the fact that Pacinian corpuscles are virtually silent in inactive joints and are activated with the onset and cessation of movement. Of course, this behavior is consistent with their rapidly adapting characteristic seen in their skin counterparts.

Golgi-Mazzoni endings (anatomically similar to Golgi tendon organs) are classified by Wyke as Type III and are located among the collagen fibers of the extrinsic and intrinsic ligaments of larger joints (e.g., knee collateral and cruciate ligaments). Along with their similarity to GTOs comes a similar functional characteristic. They are most active in their firing patterns at extremes of position in flexion/extension or other motions that stress the ligaments.

Wyke's Type IV joint receptors are primarily free nerve endings. They are located in the fibrous joint capsule, fat pads, ligaments, and the walls of blood vessels. They tend to be unmyelinated and are high threshold, nonadapting pain receptors.

As one can see, with the great variety of joint receptors there is quite a symphony of proprioceptive information coming to the central nervous system. Their primary function is in the transduction of joint position. The relative amount of information joint receptors provide when compared with muscle spindles and other mechanoreceptors is small.

A majority of joint receptors are activated at the extremes of range (i.e., situations with greater tension on the capsules or ligaments). Greater firing rates are also likely in the event of scarring of the capsule. That is to say, the joint receptors would perceive the joint to be at an extreme of range of motion before that extreme is actually achieved. Such a situation could easily arise in the case of any surgery with a capsulotomy or capsulorrhaphy. The same situation would also occur in the case of an inflammatory condition resulting in scarring of the joint capsule as in adhesive capsulitis of the shoulder or a change in the viscoelastic properties of the capsule or ligaments.[10,11]

A change in the relative tension on a joint capsule, as in adhesive capsulitis, would result in a differential level of inhibition and/or excitation between sides of the body. There is likely a difference between sides of the body based on movement pattern history, as in the case of handedness. If that relative difference were changed as a result of injury, however, the background levels of muscle tone would alter the movement patterns. Such a change has been described by Janda and is commonly observed by clinicians observing scapulohumeral rhythm. With this difference in relative levels of facilitation or inhibition, then the motor commands sent to the shoulder to perform a well-learned task (e.g., tennis serve or baseball pitch) would be inappropriate to accomplish the task or could be interfered with by spinal level reflex mechanisms. In either case, the movement patterns become dysfunctional and either introduce inaccuracy into the performance or faulty movement patterns that could result in overuse injuries.

Small Diameter Muscle Afferents (III & IV)

Another group of afferent fibers that has not received very much attention from the manual therapeutic community are the Group III and IV muscle afferents (not to be confused with Wyke's terminology for joint receptors). Group III and IV muscle afferents are thinly myelinated and unmyelinated, respectively. Group III afferents are located in the interstitial spaces of skeletal muscle either close to or within the adventitia of arterioles and venules.[12] Von Düring and Andres specifically found Group IV endings in the adventitia of small veins and even in the lymphatic vessels.[13]

The research on their reflexive effects has concentrated primarily on their effects on ventilation and circulation. In that realm, they have been found to have some very powerful effects on both systems. On the sensory input side of the reflex equation, the III and IV afferents respond to mechanical, chemical, and thermal stimuli. On the output side of the equation, the responses appear to be related to both systemic changes in respiration and cardiac output and to local changes in muscle blood flow. The former of these have been well documented in the cardio-respiratory literature and the latter in early and recent literature reviewed in Chapter 4 of this book.[14]

Clinical Implications of Small Diameter Muscle Afferents (III & IV). Considering the therapeutic techniques presented in this book and their indications, it is worthwhile to discuss mechanisms of stimulation of the Group III and IV receptors in a general sense as they relate to myofascial technique. First, Group III and IV receptors both respond to mechanical stimulation and muscular contraction. The mechanical stimulation most often used in research is that of non-noxious probing.[15] The basic conclusions have shown that more force is required to stimu-

late the receptive fields of Group III and IV than is required for muscle spindles and GTOs. The IIIs and IVs are even differentially sensitive to mechanical stimulation.[14] The second and third mechanisms are related to chemoreception and a sensitizing effect of chemicals to mechanical stimulation. These are considered as follows.

Non-noxious probing will stimulate a majority of IIIs, with noxious probing generating explosive bursts of impulses from many more IIIs. In contrast, almost all Group IV receptors require a noxious level of probing and then only a few bursts of impulses will occur. The level of probing (non-noxious and noxious) referred to in these studies exceeds the level of stimulus required by muscle spindles and GTOs by several orders of magnitude, making the forces very similar to those applied in even some of the gentlest techniques shown in this book.

Approximately half of the Group III receptors respond immediately with a vigorous increase in afferent discharge during muscular contraction. The remaining half tends to respond as the exercise session progresses (i.e., >30 seconds of isometric contraction or rhythmical contraction). This finding indicates that about half of the receptors may be more chemically rather than mechanically sensitive. Unlike the Group IIIs, Group IVs tend to rarely discharge with muscle contraction in the early stages. As the contraction progresses and is maintained, however, the frequency of their discharge increases. This finding would also support a primarily chemoreceptive role for the Group IVs.

The other studies regarding chemoreceptive properties of these two receptors have been performed with a wide variety of drugs including papervine, isoproteronol[16] (both vasodilators), bradykinins,[17] arachidonic acid[18] and lactic acid,[19] indomethacin, and aspirin.[20] The most important findings for the manual therapist relate to changes in the mechanical sensitivity of the receptors in response to some of these chemicals. Specifically, there are increases in sensitivity to mechanical stimuli with the metabolic and inflammation byproducts. On the positive side, increased concentrations of bradykinin and cyclooxygenase metabolites (both strongly associated with inflammation and injury) are likely to increase the sensitivity of Group III and IV afferents to contraction and mechanical probing. Contrary to these findings, indomethacin and aspirin, both of which decrease a muscle's ability to produce prostaglandins and thromboxanes, decreased the sensitivity of the receptors to contraction.

Given that the sensitivity of Group III and IV muscle receptors is positively affected by naturally occurring inflammation byproducts and negatively affected by anti-inflammatory drugs, the clinician needs to consider these effects during treatment. In the case of an inflammatory process (either acute or chronic), there would be a tendency for a greater increased blood flow in a muscle undergoing manipulation that was also inflamed, possibly resulting in intramuscular edema. If the patient had taken anti-inflammatory agents, however, the sensitivity to mechanical stimulation would be decreased and, therefore, the negative effects of increasing edema would be mitigated. Basically, such a line of reasoning would serve as a precaution for use of myofascial manipulation on an inflamed muscle.

To summarize, the mechanoreceptors of the mammalian body are numerous and diverse in their anatomy and function. Some are exceedingly sensitive to mechanical stimuli (e.g., Merkel's, Pacinian, Ruffini, and Meissner's corpuscles) whereas others require more vigorous stimulation (e.g., Group III and IV muscle afferents). Despite their diversity, they have two things in common. First, they are all physically connected to and perceive mechanical events from connective tissues. Second, they all provide afferent information to the central nervous system that then exhibits reflexive effects in the periphery. Some of those reflexive effects are directly motoric in nature and others are more autonomic in nature. Some of these reflexive effects will be reviewed in various levels of detail in a later section of this chapter, but no discussion of peripheral receptors is complete without

some attention to nociception and chemoreception.

Nociceptors

Nerve fibers that are selectively responsive to stimuli from damage to or that are potentially damaging to tissue are called nociceptors. They fall into three major categories: (1) mechanical, (2) thermal, or (3) polymodal, depending on the form of stimulus required. These three categories can be further classified as to their afferent nerve fibers. The thermal and mechanical stimuli are transmitted via Aδ fibers and the polymodal stimuli via the C fibers. Aδ fibers are thinly myelinated fibers that conduct impulses at 5 to 30 meters per second. C fibers are unmyelinated fibers that carry impulses at rates from 0.5 meter to 2 meters per second.

In addition to their conduction velocity characteristics and the modes of stimuli, there is another important characteristic of pain that needs to be considered. This characteristic is related heavily to chemoreception. Chemoreception is typically considered a sensory modality reserved for the tongue and nose, but in the case of pain, it becomes of extreme importance. Nociceptors demonstrate two responses to a large number of chemicals and naturally occurring agents. The chemicals either activate them or they are sensitized by them. Activation is manifested by an action potential of the nerve whereas sensitization is a lowering of the stimulus threshold required to produce an action potential. Some of the agents are included in Table 5–2.[21]

RECEPTOR INFLUENCE ON MOVEMENT

Any chapter on the neurophysiology of myofascial manipulation would be incomplete by half if it only included the afferent side of the equation. To remedy that problem, the following sections explain, in brief, a few of the reflexive and higher order sensory influences on movement and autonomic function. The explanations are concise, under the assumption that the reader

Table 5–2 Chemical or Agent Effect on Nociceptors

Chemical or Agent	Source	Effect on Nociceptors
Potassium	Damaged cells	Activation
Serotonin	Platelets	Activation
Bradykinin	Plasma kininogen	Activation
Histamine	Mast cells	Activation
Prostaglandins	Arachidonic acid-damaged cells (inflammation product)	Sensitization
Leukotrienes	Arachidonic acid-damaged cells	Sensitization
Substance P	Primary afferent	Sensitization

Source: H.L. Fields, *Pain*, p. 32, © 1987. Reproduced with permission of the McGraw-Hill Companies.

has a familiarity with these topics. If more information is desired, the reader is referred to Chapters 21–24 and 33–38 of Kandel, Schwartz, and Jessell's classic, *Principles of Neural Science, 4th Edition.*

Basics of Motor Control

Motor control is considered to be achieved through the hierarchical and sometimes parallel control processes of three different levels. The spinal cord, brain stem, and cerebral cortex each have their own independent levels of control and then work together to accomplish control. In Figure 5–6 the reader can see a relatively simple diagram of the motor system.[22] The following sections emphasize the "sensory consequences of movement upon movement" component of the model in Figure 5–6. Furthermore, some attention will be paid to the influence of myofascial pathology on the sensory consequences of movement. Unlike the sections on sensory receptors, the following sections follow a scheme of the most familiar of mechanisms moving on to some of the less familiar mechanisms and newer findings.

Muscle Stretch Reflex

Probably the best understood and most studied of the influences of peripheral receptors on

movement is that of the muscle stretch reflex (MSR), previously known as the deep tendon reflex. The MSR is a monosynaptic reflex with input from the primary and secondary endings in the muscle spindle with the major portion of the stimulus coming from the primary endings. During the MSR, the stimulus to the primary endings in the form of a sudden lengthening of the muscle is conducted by the Group Ia afferent. The Ia afferent synapses directly on an alpha motor neuron for the same muscle and excites it to the level of an action potential. This results in transmission of a motor impulse to the stimulated muscle and contraction of the muscle. All of this occurs in very short order, requiring only about 40 to 60 milliseconds.[8]

As described previously, the influence of pathology in the connective tissue can be considerable on the muscle stretch reflex. An alteration in the parallel link of the muscle spindle to its tendonous connection can occur with faulty links to other connective tissue outside of the target muscle. Connections via scarring or newly formed cross-bridges of collagen to the skin, intermuscular septa, other tendons, or even bone can occur in connective tissue pathology. Such connections could alter the MSR to either a heightened level or a lowered level of activity depending on the stimulus applied to them. In the case of pathomechanical cross-bridge formation, such an increase in the sensitivity of the

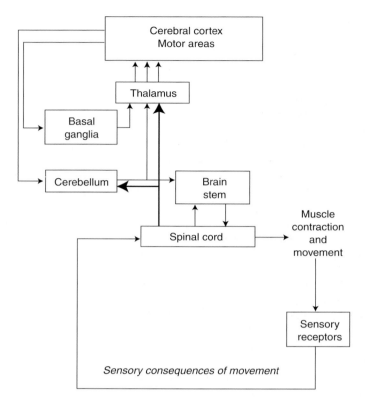

Figure 5–6 Motor system levels of control. *Source:* Reprinted with permission from J.H. Kandel et al., eds., *Principles of Neural Science*, 3rd Ed., pp. 533–547, © 1991, McGraw-Hill Companies.

MSR would alter the spinal level mechanisms of muscle tone regulation. It has been hypothesized by Janda that these changes would result in an increase of dynamic muscle tone in the agonist muscle. With changes in dynamic muscle tone and subsequent changes in movement patterns, the mechanical stresses would be different on the system resulting in connective tissue remodeling in response to Wolf's Law.

Consider an example of a patient, 3 weeks status post distal third femoral fracture with an intermedullary rod, in the supine position with the lower leg hanging over the end of a treatment mat with the knee in flexion. With the scarring that occurs, there will be adhesions between the vastus lateralis, rectus femoris, and vastus

intermedius. Each of these has a resting muscle tone. If the adhesions have formed in such a way as to differentially affect the rate of change of length in the muscles as they slide together and against each other, however, there will be a sensory mismatch. With this sensory mismatch there will also be a differential MSR response between the three muscles that was not present before the scarring occurred. This example of connective tissue pathology impact on the MSR is just one of many possible scenarios. In like manner, this example considers the impact of such a pathomechanical situation on the MSR. There are multiple other interactions to be considered, a few of which will be considered in the following section.

Golgi Tendon Organs

Golgi tendon organs, when stimulated by a change in tension, have an inhibitory effect on the agonist muscle and a facilitatory effect on the antagonist muscle. The mechanism of this event is much more complex than the MSR. In the case of the MSR, there is a monosynaptic connection of the muscle spindle afferent fibers synapsing with the alpha motor neuron for output. The afferent input from the Golgi tendon organ synapses on the Group Ib inhibitory interneurons. These interneurons receive input from multiple sources before synapsing themselves with the motor axons of either the agonist or antagonist muscles. Originally thought to be a protective mechanism to prevent tendon rupture, this same mechanism offers great utility for the manual therapist in relaxation of agonist muscle guarding and/or facilitating antagonist retraining during therapeutic exercises. The receptors for the Golgi tendon organ are specialized nodes on an axon that respond to mechanical deformation with an action potential. Therefore, the mechanical event necessary to fire the GTO does not have to be stretch; it could be direct pressure on the musculotendinous junction. The outcome regarding muscle tone is the same whether stimulated by tension or other mechanical input.

Some of the Group Ib inhibitory neurons receive converging input from Ia afferents from muscle spindles, low-threshold cutaneous afferents (e.g., Merkel's receptors and Pacinian corpuscles), joint receptors and excitatory as well as inhibitory input from several descending pathways. All of these combined inhibitory and excitatory inputs have major implications for fine motor control. The GTOs and the other inputs to the Group Ib inhibitory interneurons provide for fine control of exploratory behaviors where the amount of force being generated is critical. Therefore, the implication for these receptors' importance when learning to perform manual therapy is obvious.[23]

In addition to the implications for fine control and control in exploratory behaviors, there are ramifications for patients and their motor control. The most conspicuous of examples for problems with the GTO would be that of tendonitis. In the case of Achilles tendonitis, an inflammation of the musculotendinous junction would result in interfascicular edema inside the tendon. This edema changes the viscoelastic properties of the musculotendinous junction, resulting in a change in the tension on the braided collagen fibers that surround GTOs. Besides the change in the mechanics of the GTO, there is also the ambient change in chemical make-up of the GTO. With greater concentrations of bradykinin and cyclooxygenase metabolites (byproducts of inflammation), it is possible that the sensitivity of the GTO is increased in the same way as the sensitivity of Group III and IV muscle afferents are altered by these agents. If the GTO sensitivity were increased by inflammation byproducts, then the increased GTO firing rates would further inhibit the agonist and facilitate the antagonist and, thereby, interfere with normal motor control and movement patterns.

Joint Receptors

As previously described, joint receptors come in a variety of shapes, sizes, functional characteristics, and locations in the joints. For purposes of this discussion, we restrict our discussion to the Golgi-Mazzoni and the ligamentous free nerve endings because they are the most superficial of the joint receptors and are the most easily stimulated in the practice of myofascial manipulation. The Golgi-Mazzoni receptors are similar to Golgi tendon organs and exhibit very similar effects on motor control at a reflex level. In like manner, the free nerve endings transmit information to the spinal cord and synapse on Group Ib inhibitory interneurons. Both of these joint receptors are rapidly adapting receptors and are also known to be essentially silent in immobile joints. They are stimulated most at the extreme ranges of motion. Therefore, from a functional viewpoint, the surface of a joint capsule in which the receptors are located dictates which muscles are the agonists and which are the antagonists. In the case of the posterior capsule of the knee, rapid knee extension would result in an inhibitory effect on the quadriceps at the end of range, whereas rapid knee flexion would stimulate the joint receptors in the anterior capsule and cause inhibition of the hamstrings at the extreme of knee flexion.

Another example of the inhibitory properties of an abnormal stimulus to joint receptors is provided by Kennedy and colleagues. In their classic paper of 1982, they demonstrated that an effusion (60 cc) of the knee would result in 30% to 50% decrement in the electrical activity of the quadriceps, as measured by the Hoffman reflex, with the greatest inhibition occurring in the vastus medialis. Although they did not distinguish the particular types of receptors, they were able to show that the receptors in proximity to the joint cavity itself were very important. Under the conditions of effusion, the quadriceps were inhibited; however, when a local anesthetic was added to the effusion, the inhibition all but disappeared.[24] Clinically, these findings add even further motivation for the therapist to control joint effusion and, failing that, to make conservative recommendations for strenuous activity of the lower extremity. If such a small joint effusion can inhibit the quadriceps, then failure to control the effusion could lead to serious injury from inhibition of the surrounding musculature. One can only assume that similar findings would be seen in other diarthrodial joints with similar muscular inhibition. Such findings clearly demonstrate that a mechanical stress on the rapidly adapting receptors such as the Pacinian corpuscles is (most likely) inhibitory to quadricep motor units. Indeed, these findings offer compelling evidence that in the presence of edema or bleeding following thrust manipulation procedures, there would be a reflex inhibition of musculature surrounding that joint or related to that joint neurologically. Their findings are consistent with the prior and subsequent literature, which confirms that joint receptors are more sensitive to extremes of range. The mechanical stress placed on the joint capsule served to stimulate the joint receptors in the same manner as extremes of range of motion would.

Skin Receptors and Position Sense

The influence of skin receptors and other mechanoreceptors located in deeper tissues on motor activation levels has been documented for almost 100 years. Simple reflexes such as the flexion withdrawal reflex are spinal level systems evoked by stimulation of nociceptors. Other stimuli of a noxious nature, such as a slightly caustic agent, placed on the leg of a spinalized frog will produce the even more sophisticated movement of attempting to wipe away the stimulating agent.[25]

Hagbarth demonstrated in 1952 that a pinch stimulus to the skin of the dorsal aspect of the hind limb of a cat (i.e., opposite surface of the muscle) would inhibit the output of motor neurons to the tibialis anterior (TA) whereas the same stimulus presented to the skin on the ventral aspect (i.e., over the TA), facilitates motor neuron activity.[26] These and similar findings form the foundation for many of the facilitatory and inhibitory handling techniques employed by physical and occupational therapists today. Many of these facilitatory and/or inhibitory techniques were originated by clinicians working with neurologic clients. One common technique is that of maintained pressure over the anterior thigh, which is inhibitory to the quadriceps after an initial burst of electromyogram (EMG) activity.

The findings of changes in motor output as a result of manual contact and other stimulus input are well known. Another aspect of effects of sensory input from the skin on motor output that is not as well known is that of the contribution of skin mechanoreceptors to position sense. Psychophysical (i.e., behavioral measures of perception) such as those performed by Burgess et al and Matthews failed to demonstrate a significant deterioration of kinesthetic sense in response to anesthetizing the skin.[27,28]

The psychophysical findings would lead the clinician to think that skin mechanoreceptors have little if any influence on position sense. The work just cited, however, operates from a negative assumption. Burgess et al assumed that because elimination of skin receptors failed to negatively impact the performance of their task, that mechanoreceptors in the skin had no impact on position sense. More recent work by Collins et al investigated the threshold of perception of a muscle twitch at the wrist and found it to be attenuated by as much as 60% with voluntary movement of the same arm. An experimental manipulation of stretching the skin on the dorsum of the hand produced a 79% reduction in twitch detection threshold and a 58% reduction in position sense accuracy when compared with controls.[29]

Cohen et al performed single cell recordings from the sensory strip of the cortex in monkeys to investigate the relative effect of passive and active movement and skin stretch on cortical cell activity. They recorded from cortical cells while passively moving the monkeys' arms into flexion and abduction. They also recorded from the same cells while stretching the skin of the medial upper arm. They demonstrated that 84% of the cell recordings that responded to passive movement also responded to skin stretch. Inversely, 84% of the cells that did not respond to passive movement also did not respond to skin stretch.[30]

The findings from these two studies present fairly convincing evidence that skin mechanoreceptors serve as a source of position sense information. In addition, whether or not they contribute to the position sense information is not dependent upon external forces being applied directly to the skin.

On a functional note, the data from the Collins et al experiment indicated that, as one would expect, there was a directional bias for skin receptor firing or lack of firing. This bias produced a tendency for more phasic firing of the receptors as the monkey reached toward a right upper quadrant target with the left hand, as compared to a tonic firing of receptors as it reached toward a target in the left upper quadrant with the left hand.[29] Considering the mechanics of the situation and the extensibility of skin, these findings are not surprising.

The implication for cutaneous mechanoreceptor information functioning as position sense information is obvious. In the case of the slowly adapting receptors in the skin, a constant abnormal mechanical stimulus from an adhesion would result in an alteration in the cumulative position sense information even in the absence of motion or extremes of motion. On the other hand, an adhesion would apply forces that would easily stimulate rapidly adapting receptors when the subject engaged in either rapid or extremes of movement. These same receptors, as cited previously, have either facilitating or inhibitory effects on muscles that are agonists or antagonists for movements normally associated with input from that area of skin.

Nociceptors

The influence of pain on movement and control is probably the most obvious sign of change in movement control that is brought on by peripheral stimulation. The initial response to acute pain from mechanical or thermal stimuli is transmitted to the nervous system via Aδ fibers, which are myelinated rapidly conducting fibers. The withdrawal reflex seen in even spinalized animals is very fast. The response to pain of a more polymodal nature, such as that mediated by inflammation byproducts and other neuroactive agents, is typically more complicated.

Polymodal pain is carried on C fibers, which are slowly conducting fibers. Polymodal pain, most likely because of its continuing nature, also has the characteristic of inducing some type of behavioral response. This response may be at a spinal level or virtually any other level of the nervous system. The responses can vary from obvious muscle guarding in the surrounding musculature to help decrease movement, to inhibition of a muscle to decrease movement that would increase the pain.

One misconception that has been propagated from one author to the next is that of the "pain-spasm-pain" cycle. Unfortunately, as reviewed so eloquently by Walsh,[1] and Simons and Mense,[31] this misconception is based on a misunderstanding of the involved motor reflexes. The original hypothesis stated that pain increased γ-motor neuron activity, which would stimulate or increase the sensitivity of the muscle spindle and result in an increased α-motor neuron activity and muscle contraction. The major problem with this theory is the fact that muscle pain does not result in increased EMG activity. Furthermore, the timing and intensity of the EMG activity does not correlate with the reported levels of pain.

These findings present a contradiction to the practitioner of manual therapy and any acute observer of posture and movement. It is relatively easy to identify muscle asymmetries in bulk as well as in muscular activity during movement. If muscular pain and the apparent increases in muscle tone are not caused by spasm, then what does cause the increase in muscle tone? This question is actually two-fold. First, what does pain have to do with increased muscle tone? Second, what is muscle tone/spasm? The following sections explain some of the current thinking regarding these topics. Since we need a definition or explanation of muscle tone, that section is presented first.

MUSCLE TONE

Muscle tone (taken from the Greek τεινω, [teino]) has long been associated with muscular contraction; however, it is actually more complex than just a contraction. Certainly, a typical muscle contraction or level of muscle tone, as understood from the sliding filament theory, cannot occur without an electrical action potential. Since, as cited previously, an action potential is not typically discernible via EMG in the case of muscle spasm or even "normal" resting muscle tone, it must entail more than just an electrogenic activation of the actinomyosin complex. Simons and Mense have offered an excellent review of muscle tone and its relation to clinical muscle pain. In it, they describe muscle tone as consisting of two types of muscle tone. The first one is known as electrogenic tone and the second one is viscoelastic tone (Figure 5–7).

Electrogenic Muscle Tone

Electrogenic tone can be categorized into three levels. The first level is resting muscle

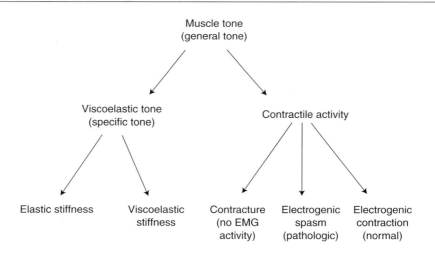

Figure 5–7 Muscle tone. *Source:* Adapted from Understanding and Measurement of Muscle Tone as Related to Clinical Muscle Pain, *Pain*, No. 75 (pp. 1–17) by D.G. Simons and S. Mense with permission of W.B. Saunders Company, © 1998.

tone. This muscle tone has historically been explained as a postural low-level tonic discharge of motor neurons. As explained by Walsh, this misconception was begun by Waller and was based on a generally inapplicable experiment reported by Brondegeest in 1860. Waller, and later the Sherrington school, explained resting muscle tone with the muscle stretch reflex. Such an explanation would, by definition, require an action potential to be generated in α-motor neurons. Activation of α-motor neurons would activate motor units, which would be perceptible by EMG. All efforts to document resting muscle tone via EMG have failed.[32–34] This is not to deny that some form of contracture is occurring in the muscle. Physiologists tend to define contracture as an endogenous shortening of the muscular contractile apparatus in the absence of EMG activity initiated by anterior horn cells.[35] With this definition, there are cross-bridges formed but they have not resulted from an action potential from the myoneural junction.

The second level of electrogenic muscle tone is what Simons and Mense refer to as electrogenic spasm. This particular type of contraction is an involuntary contraction that is directly associated with measurement EMG activity from that muscle.[31] Voluntary muscular contraction is the third and last level of muscle tone and requires no explanation.

Before we move on to a more in-depth explanation of recent findings regarding viscoelastic tone, it is useful to discuss in a little more depth ideas related to clinical muscle spasm. As we have already stated, a pain-spasm-pain cycle is an insupportable hypothesis in the sense of an electrogenic spasm. As anyone who has worked on another human or even mammal will attest, however, pain and changes in compressibility of muscular tissue are easily discernible by palpation. In this regard, the findings related to trigger points and tension-type headache (T-TH) are particularly revealing. In T-TH it is easy to palpate taut bands of muscle. These bands, while often associated with trigger points, do not demonstrate observable EMG activity. The trigger points, themselves, however, have been

shown to demonstrate localized electrical activity in the confined area of the trigger point.[36] It appears that these taut bands of muscle are the result of the same contracture mechanisms described by physiologists.

Other forms of muscle contraction of particular interest to clinicians fall into two categories. The first form we know as involuntary guarding, where there is unnecessary muscular contraction that limits movement. The second form could best be described as inefficient use. Most clinicians are aware that because of guarding, pain avoidance, and other causes, patients will move in manners that are inefficient. These inefficient movement patterns can have serious consequences. Consider, for example, a marathon runner who gets a blister over the head of the fifth metatarsal at mile 3 of the race. Such a minor injury has been known to have consequences of a femoral head stress fracture by the end of the marathon. The same such inefficient use can occur with painful muscles and/or trigger points. Lack of appropriate relaxation between contractions of the upper trapezius has been demonstrated by Elert et al, and Ivanichev demonstrated that muscles with trigger points failed to relax appropriately during alternating movements as they became fatigued.[37,38]

An understanding of these various levels of muscle tone, which are associated with electrical activity in the muscle, is certainly important for the clinician. Also, an insight into the influences of various receptors on α-motor neuron and γ-motor neuron activity is useful for understanding control; however, this volume relates more specifically the manipulation of myofascial tissues. Consequently, the next section on viscoelasticity is very important and will help the reader to understand some of the very rapid results seen with myofascial manipulation.

Viscoelastic Muscle Tone

The viscoelastic muscle tone, or specific tone, is made up of an elastic component and a viscoelastic component. The purely elastic component, by definition, requires a steady force to produce a deformation of the substance, which in this case is myofascial connective tissue. As we know, the collagen and other structural proteins of myofascial tissue are not the only components of connective tissue. These tissues also contain various other proteins in addition to their obvious structural systems. These other substances are primarily in fluid form and have varying degrees of viscosity or "fluid stiffness." The primary component of noncontractile fluid component is water, which is retained by the nonsulfated glycosaminoglycans (GAGs) and makes up about 70% of the extracellular matrix. The second component is the sulfated version of GAGs, which account for the tissue cohesiveness. Another fluid component of myofascial tissue is actin. Although actin certainly comprises a large complement of muscle itself, it is also abundantly present in noncontractile fluid and serves cell motility and intracellular structure functions. This protein is actually fluid in its purified form and, much like syrup, will form strings when picked up on a glass rod or other stirring device.

The GAGs, actin, and myosin all contribute to the viscoelasticity of myofascial tissue. Unlike elasticity, the stiffness of viscoelasticity is velocity dependent. Also, it is worthy of note that unlike the velocity dependence of spasticity, the relationship between viscoelasticity and velocity of movement is purely mechanical. The mechanical viscoelasticity characteristic and the structural elasticity of the structural proteins combine to make up the specific tone of a muscle that is unrelated to contractile activity.

Viscoelasticity of muscle, or viscoelastic tone, affects movement and postural control. The sensation(s) from the musculoskeletal system that prompt mammals to stretch after remaining still are relatively undefined concerning their sensory mechanisms. Concerning posture, there are mechanical properties of muscle (largely unexplored until recently) that tend to support a resting stiffness of muscles in posturally supported humans that is unrelated to EMG activity with the exception of occasional corrective bursts of activity. The properties of myofascial

tissues that prompt the stretching behavior and account for maintenance of static balance, however, have experienced an abundance of study over the past 10 years and a new flurry of activity during 1998 and 1999. This "new" property is known as *thixotropy*.

Thixotropy

Defined

Thixotropy [θιξισ (touch) and τροπη (turning or change)], as a term, is new to many people across the entire spectrum of clinicians who use manual therapeutics. It is not, however, new to physiologists involved in the study of muscle and tissue mechanics. Thixotropy describes a state of stiffness of a fluid that is dependent on the past history of movement. There are a number of common substances that exhibit thixotropy. Tomato catsup is probably the most common. After sitting in the bottle, catsup becomes very stiff and difficult to get out of the bottle. With just a little stirring, the stiffness decreases substantially.[39]

Thixotropy is a *physical property of muscle* and other tissues and not a response to some neurophysiologic event. The mere act of moving a substance with thixotropic properties will result in a reduction of stiffness. The reverse is also true, if a thixotropic substance remains still for a given period of time (variable dependent upon the substance), the substance will become stiffer.

In order to measure thixotropy, physiologists have used torque motors with very small torques of approximately 0.1 Newton.meters (Nm). Under conditions of a sinusoidal motion of the wrist, the amplitude of a motion of the wrist is about 0.02 radians (1.14°). With a movement of the wrist in an amplitude of approximately .075 radians for only three cycles, the amplitude of the passive wrist movement with the same 0.1 Nm of torque increases to about 0.06 radians (3.42°). These amplitudes are very small so as to avoid stirring the muscle; however, it is important to note that a brief interruption of as little as 2.5 seconds returned stiffness to its original levels. Also of note is the fact that this stiffness, which is restorable in as little as 2.5 seconds, is possible at most any length with the exception of a position of extreme stretch. As one can tell, the amount of stirring and the amount of interruption of motion can be very small. Now that we understand the basic property and concept, we delve into the mechanisms.[40]

Possible Mechanisms of Thixotropy in Muscle

Several investigators have hypothesized that the thixotropic properties of muscle originate at the cross-bridge mechanisms in muscle. Campbell and Lakie have proposed that the thixotropic behavior of relaxed skeletal muscle may be explained by a tendency for some of the cross-bridges to connect even in the absence of an action potential. As described by Hill, the early stage of the tension response to movement appears to be dependent on the duration of the rest period (no movement) and the filamentary release tension, which occurs later in the movement and is linked to the stretch velocity.[41] Campbell and Lakie summarize their explanation of thixotropy, which they attribute to a model of undetached cross-bridge mechanisms, by saying, "The molecular motors of muscle may be idling rather than switched off when the muscle is relaxed."[42(p957)]

There is another hypothesis that can explain the thixotropy of muscle. This hypothesis put forward by Mutungi and Ranatunga[43] and other investigators would attribute the viscoelastic properties of relaxed skeletal muscle to titin filaments. Titin filaments are exceptionally large structural proteins in muscle, which link the thick myosin filaments to the Z-lines of muscle. Titin filaments tend to adopt a random-coil configuration when relaxed and that uncoils with stretching. Consequently, titin does not offer a very viable explanation for thixotropy of muscle but with its increase in tension at extremes of range, it may contribute to the resistance felt in muscle when it is stretched to near its limits of range of motion.[43,44]

Clinical Implications of Thixotropy

Considering the ranges of motion used in measurement of thixotropy, it is questionable whether thixotropy has any practical application to clinical practice. This author proposes that thixotropy may offer an explanation for the phenomenon of palpable "muscle spasms" that are found on examination of patients with trigger points. As previously cited, highly localized electrical activity has been found in trigger points. These same trigger points have also been identified by Simons[45] as corresponding anatomically with the intramuscular portion of motor nerve terminals. It is possible that the localized electrical activity is adequate to sensitize nociceptors in the area of a trigger point. The nociceptive agents released may also destabilize the T-tubules enough to result in a higher calcium concentration within myofibrils. This would result in a larger number of cross-bridges being formed between the myosin heads and troponin, which would increase the stiffness (i.e., thixotropy). Such an increase in cross-bridges would decrease the pliability of muscle in the immediately surrounding muscle tissue. This phenomenon is feasible to explain the experience of deep massage being beneficial to increase the pliability of the muscle around trigger points. According to the "pain-spasm-pain" cycle, deep massage of a trigger point should increase the pain and, therefore, the spasm, with even more pain. This does not always occur in practice, as many practitioners can testify that deep massage can "decrease the spasm."

Neurophysiological Implications of Thixotropy

The mechanical properties of thixotropy have been reviewed in the previous sections. These properties, obviously, apply to the largest complement of muscle, the extrafusal fibers. Extrafusal fibers are only part of the picture, however. As Proske et al have reviewed, thixotropy, as a mechanical property, has a profound influence on muscle spindles and their afferent neurons.[39] These influences are too numerous to review here, but the resting discharge of primary spindle afferents and their sensitivity to muscle stretch are dependent on the previous history of movements and/or contraction. In several research studies, it has been demonstrated that when a conditioning movement or contraction is performed, such as an isometric contraction in the shortened position, the afferent discharge from muscle spindles is increased.[8] The reverse is observed in an isometric contraction in the lengthened position. This phenomenon is not a facilitation of the spinal cord mechanisms but rather a sensitization or desensitization, as the case may be, of the muscle spindle. Studies of this phenomenon using a mechanically stimulated muscle stretch reflex (i.e., tendon tap) produce the phenomenon; however, similar studies with the Hoffman reflex (i.e., an electrophysiological analog of the tendon tap) have failed to show the same results.

Another potential influence of thixotropy can be postulated based on the biochemical, biophysical, and neurophysiological properties of joint capsules and other connective tissues. First, the biochemistry and biophysics of the sulfated GAGs have shown them to be responsible for the cohesiveness of connective tissue. Second, with this increased cohesiveness comes an increased initial resistance to active or passive stretch. Consequently, one would expect an initial afferent discharge from joint receptors after a joint has remained still for a few minutes. Indeed, Walsh and Wright demonstrated that thixotropy occurs at the human hip, with the amplitude of the resonant frequency of a sinusoidally abducting/adducting hip almost doubling in response to a stirring motion of large amplitude.[46] Whether this thixotropic resistance to initial movement actually produces an increased afferent discharge very early in the time course of the movement remains to be tested. Nevertheless, if the fluid mechanics of a joint capsule, musculotendinous junction, or direct muscular attachment to bone were changed by inflammation byproducts, then the afferent output from those receptors could certainly be either increased or decreased. Such an event may explain

some of the faulty movement patterns or holding patterns described by Janda and Feldenkrais.

We have reviewed the basic receptor anatomy and physiology for most of the somatosensory system with the exception of the vestibular system. We have also reviewed some of the interactions of the somatosensory system with the motor output system with particular emphasis on that portion related to the myofascial system. Now that we have finished the neuromechanical background for myofascial manipulation, we move into some direct application of this physiology and biophysics.

APPLICATION TO SPECIFIC THERAPEUTIC TECHNIQUES

The following sections are designed to outline examples of specific application of the science heretofore presented. This application takes the form of (1) a very brief discussion of the particular technique to which application is made; (2) a discussion of the pathology/pathomechanics addressed by the particular technique; (3) a proposed theoretical mechanism, whereby these techniques may influence the somatosensory system; and (4) proposed mechanisms for alterations in motor control are engendered by the technique under consideration.

Anterolateral Fascial Elongation

The anterolateral fascial elongation technique (Figures 8–96 and 8–97) is useful to consider as its logic and associated neuromechanical characteristics can apply to virtually all of the superficial techniques described in this book. The anterior lateral fascial elongation technique, as described later in this book, primarily stretches the superficial fascial sheath in a diagonal pattern across the anterior surface of the body. In doing so, the technique is designed to treat a number of restrictions at varying levels. At the interface between the skin and the superficial fascia, there may be restrictions secondary to blunt trauma and scarring. In the superficial fascia itself and its interface with the pectoralis major and the external oblique abdominals, the sheath is continuous from the proximal humerus, clavicle, and anterior shoulder down to the contralateral crest of the ilium, thoracolumbar fascia, anterior superior iliac spine, inguinal ligament, and the pubis.

Restrictions of the superficial fascia of the anterior trunk have mechanical implications for posture and virtually all movements of the trunk and upper and lower extremities. Certainly, there are mechanical restrictions of mobility but given that patients develop such faulty postural habits, the pathomechanical implications for the body as a whole are most likely seated in position sense. Restrictions in the superficial fascia would result in a continuous and abnormal stimulus of the slowly adapting mechanoreceptors in the skin and all the succeeding layers of the superficial fascia. Because the mechanical restriction in the skin and superficial fascia is very similar to that found in the experiment performed by Cohen et al, some direct postulates are in order.

Cohen and colleagues found increased activity of somatosensory cortical cells representing skin receptive fields in the axilla and the skin of the medial proximal arm associated with parallel skin stretching, passive movement, and active movement. They were able to demonstrate this same highly correlated activity in a variety of tasks including reaction time tasks, holding tasks, and active movement of the arm. The shortened range of skin produced very little activity in tactile receptors of the axilla and upper arm. This is in contrast to movements into shoulder flexion or shoulder flexion with abduction, which increased the activity.[46] Furthermore, the greater the stretch in either amplitude or movement, the greater the firing rate of phasic (rapidly adapting) receptors (e.g., Pacinian corpuscles).

These findings are completely logical and intuitive when one considers human postural phenomena observed by clinicians. Consider a patient who is 3 to 4 weeks post cholecystectomy via a left upper quadrant incision rather than a laproscopic procedure. A phasic stimulus of skin receptors during erect sitting or right shoulder

flexion would be perceived as a "greater than resting or normal position" burst of activity. In that case, the patient would return to a position that was more in line with resting position. If a mechanical restriction resulted in an abnormal phasic stimulus or tonic stimulus, then the interpretation by the system would be that the patient was in a stretched position when, in fact, the position might be neutral. Consequently, the patient would tend to move into a position that decreases the firing activity of the phasic and/or tonic receptors. This position is then perceived, via the skin receptive fields, as normal and further shortening of the superficial fascia occurs. This faulty receptor activity and the position sense activity it provides soon becomes the basis for postural perception.

Historically, the theoretical basis for such behavior has been that of pain avoidance. Certainly pain avoidance behavior is a reasonable and patent argument in the early stages but after several weeks of healing, the pain disappears. What remains is the new position sense reference from skin and superficial fascia receptors.

Another hypothesis concerning the continued behavior of avoiding elongation is that of alteration in motor programs (motor memories) to fit the new and dysfunctional behavior. Considering the amount of practice required to change a very well learned motor program, this is not likely. Consider, for example, attempting to change one's signature. It is possible, but on a practical level, it is not probable secondary to the huge volume (millions of repetitions) of practice required. It is very likely that this new position sense stimulus from the skin rapidly adapting and slowly adapting receptors function in an inhibitory fashion just like their Golgi tendon organ and Golgi-Mazzoni type joint receptors, by inhibiting muscles which would further stretch these receptors.

Such a postulate is based on the findings of numerous investigators of the inhibitory influences of GTOs and joint receptors on motor output. It is also in agreement with Janda's model of altered muscle function and motor performance resulting from "inadequate proprioceptive stimuli," which is probably more correctly stated as inappropriate or mismatched proprioceptive stimuli.[30] One exception is noted, and that is that the just logic described cannot validly be applied to the Bindegwebsmassage type of stroke or the skin rolling. This is because their goals and physiology are not directly connected to the evidence supplied by Cohen et al.

Iliac Crest Release Technique

This technique is useful to consider, as it is a moderately deep technique (Figures 8–20A, B, and 8–21). As described, it is executed by applying an anterior directed force through the fingers moving from the border of the iliac crest anteriorly and superiorly on to the thoracolumbar fascia and the insertion of the deep erector spinae and quadratus lumborum. This particular technique addresses restriction of the thoracolumbar fascia and the muscular and ligamentous attachments. Bogduk and Macintosh discussed the anatomy of the thoracolumbar fascia with its two layers connecting to the crest of the ilium. This anatomy makes its mechanics somewhat complicated and allows it to contribute to stabilization of the spine in virtually all movements, with the possible exception of side bending to the same side.[47]

There is some disagreement concerning the density of mechanoreceptors in the thoracolumbar fascia. An assumption that the connective tissue in this structure is no different from that found in the shoulder, knee, and ankle would lead one to conclude that the receptors consist of free nerve endings, Pacinian corpuscles, Ruffini endings, Golgi-Mazzoni endings, and others associated with ligamentous structure. In a study by Bednar et al, however, they failed to find a significant density of mechanoreceptors in the thoracolumbar fascia of patients with chronic back pain.[48] They concluded that there were differences in the density of receptors between normal subjects and persons with back pain. One major caveat concerning this study is that it was performed with standard histologic staining and no specifics were noted concerning the area

of the thoracolumbar fascia from which samples were taken during surgery. Yahia et al found Ruffini endings and Vata-Pacini corpuscles (a specialized form of Pacinian corpuscles). These samples were also taken from surgical patients. Also, Yahia's samples were prepared with immunohistochemical staining techniques that targeted neural filament protein.[49]

With the documented presence of Ruffini endings and Pacinian-like corpuscles in the thoracolumbar fascia, it is likely that a restriction of the thoracolumbar fascia would produce an abnormal afferent stimulus. This abnormal stimulus from normal motions or positions would result in an abnormally excited or inhibited level of activity for the motor units of the abdominal, paraspinal, and quadratus lumborum musculature. Therefore, decreasing restrictions with the iliac crest release technique would help to correct this abnormal afferent outflow. Such a correction would allow the relative levels of excitation and inhibitions to return to levels dictated by the normal motor programs as opposed to inappropriate proprioceptive signals.

Diaphragmatic Techniques

Techniques for correcting restrictions in the diaphragm and inferior border of the rib cage are important in facilitating patients having difficulty with postural reeducation. The techniques progress from a gentle technique of stretching the superficial to middle layer restrictions just inferior to the anterior rib cage to those that involve grasping the inferior portion of the rib cage, in a seated position, and stretching it superiorly and anteriorly while asking the patient to sit up straighter and inhale deeply (Figures 8–43, 8–44, 8–45, and 8–46).

These techniques address restrictions that are very deep in the thoracic and abdominal cavities. Although directly addressing restrictions in the thoracic cavity is not possible, it is possible to affect restrictions in the mediastinum by stretching the diaphragm and deep fascia of the abdomen and diaphragm. Such restrictions can lead to or be the result of multiple postural problems such as forward-head, protracted shoulders, and a general slumped posture in sitting.

The pathomechanics of slumped posture and forward-head are fairly well understood. With an increasingly forward-head posture comes a tendency for the ribs and sternum to move inferiorly and posteriorly. This leads to a shortening of the connective tissue in the abdomen and in the thorax. With decreasing length comes a tendency for increased afferent activity from the tension receptors (e.g., GTOs, Rufinni endings, and Pacinian corpuscles). An increase in tension on these receptors, more especially the GTOs, of the central tendon of the diaphragm has been shown to elicit a complex inhibitory effect on the external intercostals and the diaphragm.[50] All of this inhibitory activity results in a reduction in lung volume. Over time, as lung capacity is diminished by these inhibitory processes, the connective tissue would remodel to its new length resulting in a new "set" for the normal tension on the tendon. The manipulation techniques described herein allow for a lengthening of the diaphragm along its anterior borders with a resultant, postulated reduction in the inhibitory activity of the GTOs.

Transverse Muscle Bend of the Erector Spinae

This technique is relatively simple to perform and depending upon the vigor with which it is done can have exclusively neurophysiologic effects or, performed more aggressively, can have mechanical effects. The technique is basically one of bending the muscle as if bending a garden hose (Figures 8–15 and 8–16A, B). The technique can also be modified as in the quadriceps and hamstring technique to include bending and some muscle rolling and lifting actions. No matter what particular technique is used, the result is a multidirectional mechanical stress with the least emphasis on longitudinal stretching.

The major benefit of the technique appears to be in improving the mobility of muscle on muscle and of individual muscle fascicles on other muscle fascicles. If a restriction occurs between two fascicles or two muscles, then the resulting altered mechanics produces a sensory mismatch and inappropriate proprioception from the muscle. Such a case has been described previously in the section on implications of muscle spindles and GTOs.

While the influence of intermuscular and/or interfascicular adhesions on afferent and efferent neural activity are fairly common knowledge among therapists, there are other pathological problems related to such adhesions and benefits related to a transverse muscle play technique.

The pathology of such adhesions and, more importantly, the changes in intramuscular pressure caused by them relates to influences of thixotropy and the Group III and IV afferents. Adhesions of such a nature can lead to a local irritation of the muscle and a destabilization of the cell membrane adequate to cause a release of calcium into the myofibrils. This release of calcium will result in the formation of cross-bridges without benefit of an action potential and increased thixotropic resistance to stretch. Second, increases in intramuscular pressure have been directly associated with increased afferent action potentials of the Group III and IV afferents coming from the arterioles, venules, and connective tissue in proximity to these structures. Such afferent activity results in cardiovascular and pulmonary changes on a systemic level and an autonomic response of increased blood flow at a local level.

The treatment techniques themselves also have direct effects on the thixotropy of the system and the Group III and IV afferents. The muscle play motion of the muscle would provide a mechanical stimulus to aid in decreasing the thixotropic resistance to motion. Next, the technique would have direct effects on the Group III afferents, with resultant changes in local blood flow and systemic cardiovascular and pulmonary effects. The changes in thixotropy engendered by the techniques most likely also extend to changing the outflow from the muscle spindles themselves with all the cascade of effects from them.

CONCLUSION

Much of the material presented in the early sections of this chapter may appear to be weighted heavily toward basic science. It is highly probable, however, that a significant part of benefit derived from the techniques is neurophysiological in origin due to the rapidity of their effects and the relatively longer period of time required for remodeling. A number of these techniques can be viewed as methods to prepare the patient to be able to function in a manner that will lead to more functional remodeling of collagen.

We have endeavored to explain and expound, for the clinician, the relevant issues of mechanoreceptor anatomy and physiology. Moreover, we have summarized some of the recent findings of the influence of skin and joint receptors on position sense and myofascial tone. Later, in an effort to elucidate some of the more recent literature, we discussed concepts of thixotropy and their importance in muscle tone. Finally, we have attempted to connect the science directly to the techniques in this volume proceeding from the superficial to the deeper techniques.

The practitioner is encouraged to apply the science and neurophysiology where valid but to be cautious in extending their explanation too far afield from the intent of the science. Furthermore, the practitioner should remember that many manual techniques appear to have no rational explanation but appear to consistently benefit the patient. Consequently, the practitioner should use the science for explanation, when they can, while continuing to use the art of manual therapy to heal and always continue to investigate the explanations for the effects seen.

REFERENCES

1. Walsh EG. *Muscles, Masses and Motion. The Physiology of Normality, Hypotonicity, Spasticity and Rigidity.* New York: Cambridge University Press; 1992.

2. Gardner EP, Martin JH, Jessell TM. In: ER Kandel, JH Schwartz, TM Jessell, eds. *Principles of Neural Science,* 4th ed. New York: McGraw-Hill; 2000:430–449.

3. Hagen-Torn O. Entwicklung und Bau der Synoviamembranen. *Arch Mikros Anat.*1882; 21:591–663.

4. Gardner E. Nerve terminals associated with the knee joint of the mouse. *Anat Rec.* 1942;83:401–419.

5. Zimny ML, Schutte M, Dabezies E. Mechanoreceptors in the human anterior cruciate ligament. *Anat Rec.* 1986: 214;204–209.

6. Zimny ML. Mechanoreceptors in articular tissues. *Am J Anat.* 1988;182:16–32.

7. Basbaum AI, Jessell TM. The perception of pain. In: ER Kandel, JH Schwartz, TM Jessell, eds. *Principles of Neural Science,* 4th ed. New York: McGraw-Hill; 2000:472–491.

8. Pearson K, Gordon J. Spinal reflexes. In: ER Kandel, JH Schwartz, TM Jessell, eds. *Principles of Neural Science,* 4th ed. New York: McGraw-Hill; 2000:713–736.

9. Houk J, Henneman E. Responses of Golgi tendon organs to active contractions of the soleus muscle of the cat. *J Neurophysiol.* 1967;30:466–481.

10. Wyke B. The neurology of joints. *Ann Royal Coll Surg Eng.* 1967;41:25–50.

11. Zimny ML. Mechanoreceptors in articular tissues. *Am J Anat.* 1988;182:16–32.

12. Stacey MJ. Free nerve endings in skeletal muscle of the cat. *J Anat.* 1969;105:231–254.

13. Von Düring M, Andres KH. Topography and ultrastructure of group III and IV nerve terminals of cat gastrocnemius-soleus muscle. In: W Zenker, WL Neuhuber, eds. *The Primary Afferent Neuron: A Survey of Recent Morpho-Functional Aspects.* New York: Plenum; 1990:35–41.

14. Kaufman MP. Afferents from limb skeletal muscle. In: JA Dempsey, AI Pack, eds. *Regulation of Breathing,* 2nd ed. New York: Marcel Dekker; 1995:583–617.

15. Kumazawa TN, Mizumura K. Thin-fibre receptors responding to mechanical, chemical and thermal stimulation in the skeletal muscle of the dog. *J Physiol.* 1977;273:179–194.

16. Haouzi P, Hill JM, Lewis BK, Kaufman MP. Responses of group III and IV muscle afferents to distension of the peripheral vascular bed. *J Appl Physiol.* 1999;87: 545–553.

17. Mense S. Nervous outflow from skeletal muscle following chemical noxious stimulation. *J Physiol.* 1977;267: 75–88.

18. Rotto DM, Schultz HD, Longhurst JC, Kaufman MP. Sensitization of group III muscle afferents to static contraction by products of arachidonic acid metabolism. *J Appl Physiol.* 1990;68:861–867.

19. Sinoway LI, Hill JM, Pickar JG, Kaufman MP. Effects of contraction and lactic acid on discharge of group III muscle afferents in cats. *J Neurophysiol.* 1993;69: 1053–1059.

20. Rotto DM, Hill JM, Schultz HD, Kaufman MP. Cyclooxygenase blockade attenuates the responses of group IV muscle afferents to static contraction. *Am J Physiol.* 1990;259:H745–H750.

21. Fields HL. *Pain.* New York: McGraw-Hill; 1987.

22. Ghez C. The control of movement. In: ER Kandel, JH Schwartz, TM Jessell, eds. *Principles of Neural Science,* 3rd ed. New York: Appleton & Lange; 1991:533–547.

23. Houk J, Crago PE, Rymer WZ. Functional properties of the Golgi tendon organs. In: *Spinal and Supraspinal Mechanisms of Voluntary Motor Control and Locomotions, Vol. 8, Progress in Clinical Neurophysiology.* Basel: Karger; 1980:33–43.

24. Kennedy JC, Alexander IJ, Hayes KC. Nerve supply of the human knee and its functional importance. *Am J Sports Med.* 1982;10:329–335.

25. Berkinblit MB, Feldman AG, Fukson OI. Adaptability of innate motor patterns and motor control mechanisms. *Behav Brain Sci.* 1986;9:585–638.

26. Hagbarth KE. Excitatory and inhibitory skin areas for flexor and extensor notoneurones. *Acta Physiol Scan.* 1952;26:1–58.

27. Burgess PR, Wei JY, Clark FJ, Simon J. Signaling of kinesthetic information by peripheral sensory receptors. *Annl Rev Neurosci.* 1982;5:171–187.

28. Matthews PB. Proprioceptors and their contribution to somatosensory mapping: complex messages require complex processing. *Can J Physiol Pharmacol.* 1988;66: 430–438.

29. Collins DF, Cameron T, Gillard DM, Prochazka A. Muscular sense is attenuated when humans move. *J Physiol.* 1998; 508:635–643.

30. Cohen DAD, Prud'homme MJL, Kalaska JF. Tactile activity in primate cortex during active arm movements: Correlation with receptive field properties. *J Neurophysiol.*1994;71:161–172.

31. Simons DG, Mense S. Understanding and measurement of muscle tone as related to clinical muscle pain. *Pain.* 1998;75:1–17.

32. Clemmesen S. Some studies on muscle tone. *Proc R Soc Med.* 1951;44:637–646.

33. Ralston HJ, Libet B. The question of tonus in skeletal muscle. *Am J Phys Med.* 1953;32:85–92.

34. Basmajian JV. New views on muscular tone and relaxation. *Can Med Assoc J.* 1957;77: 203–205.

35. DiMauro S, Tsujino S. Non-lysosomal glycogenoses. In: AG Engel, C Franzini-Armstrong, eds. *Myology,* 2nd ed., vol. 2. McGraw-Hill; 1994:1554–1576.

36. Simons DG, Hong CZ, Simons LS. Prevalence of spontaneous electrical activity at trigger spots and control sites in rabbit muscle. *J Musculoskelet Pain.* 1995;3: 35–48.

37. Elert J, Dahlqvist SR, Almay B, Eisemann M. Muscle endurance, muscle tension and personality traits in patients with muscle or joint pain: A pilot study. *J Rheumatol.* 1993;20:1550–1556.

38. Ivanichev GA. *Painful Muscle Hypertonus* (in Russian). Kazan: Kazan University Press; 1990.

39. Proske U, Morgan DL, Gregory JE. Thixotropy in skeletal muscle and in muscle spindles: A review. *Prog Neurobiol* 1993;41:705–721.

40. Walsh EG. *Muscles, Masses and Motion. The Physiology of Normality, Hypotonicity, Spasticity and Rigidity.* New York: Cambridge University Press; 1992.

41. Hill DK. Tension due to interaction between the sliding filaments in resting striated muscle. The effect of stimulation. *J Physiol.* 1968;199:637–684.

42. Campbell KS, Lakie M. A cross-bridge mechanism can explain the thixotropic short-range elastic component of relaxed frog skeletal muscle. *J Physiol.* 1998;510: 941–962.

43. Mutungi G, Ranatunga KW. The viscous, viscoelastic and elastic characteristics of resting fast and slow mammalian (rat) muscle fibres. *J Physiol.* 1996;496:827–836.

44. Linke WA, Bartoo ML, Ivemeyer M, Bollack GH. Limits of titin extension in single cardiac myofibrils. *J Musc Res Cell Motil.* 1996;17:425–438.

45. Simons DG. Clinical and etiological update on myofascial pain due to trigger points. *J Musculoskel Pain.* 1996;4:93–121.

46. Walsh EG, Wright GW. Postural thixotropy at the human hip. *Q J Exp Physiol.* 1988;73:369–377.

47. Jull GA, Janda V. Muscles and motor control in low back pain: Assessment and management. In: LT Twomey, JR Taylor, eds. *Physical Therapy of the Low Back.* New York: Churchill Livingstone; 1987.

48. Bogduk N, Macintosh JE. The applied anatomy of the thoracolumbar fascia. *Spine.* 1984;9(2):164–170.

49. Bednar DA, Orr FW, Simon GT. Observations on the pathomorphology of the thoracolumbar fascia in chronic mechanical back pain. A microscopic study. *Spine.* 1995; 20(10):1161–1164.

50. Yahia L, Rhalmi S, Newman N, Isler M. Sensory innervation of human thoracolumbar fascia. An immunohistochemical study. *Acta Orthop Scand.* 1992; 63(2):195–197.

CHAPTER 6

Muscle Pain Syndromes

Jan Dommerholt

Muscle pain syndromes are being diagnosed today using specific criteria, and patients with these conditions are increasingly being referred to physical therapists for evaluation and treatment. Physical therapists need to understand the nature of these syndromes, how patients with these syndromes are best rehabilitated, and how myofascial manipulation fits into the rehabilitation program. Historically, pain from muscles has been described in multiple terms, including fibrositis, myofasciitis, muscular rheumatism, rheumatic myositis, muscle hardening, myogelosis, myofascial pain, and myalgia.[1] Any of these terms has been associated with examinations of patients who had pain of unknown etiology, questionable dysfunction, or negative diagnostic workups. In 1816, Balfour reported "patients as having a large number of nodular tumours and thickenings which were painful to the touch, and from which pains shot to neighbouring parts."[2] In 1904, Stockman described "chronic rheumatism" as characterized by "fibrous indurations [that are] more defined and circumscribed, varying in size from a small-shot or split-pea to

an almond, or even half a walnut.... Very frequently the thickening takes the form of a strand or cord running through the fascia or subcutaneous tissue."[2] Similar concepts, referred to as "muscle hardening" and "myogelosis" appeared in the German literature in 1921 and 1931 respectively.[3,4] In a recent review, Simons postulated that the concept of myogelosis is virtually identical to the concept of trigger points, a term introduced in 1942 by Travell and colleagues with the addition of "myofascial" in 1952.[5–7] The term "fibrositis" was first coined by Gowers in 1904.[8] For many years, persons with fibrositis syndrome were thought to have characteristic tender nodules, however, without an identified histopathological basis. It was not until the late 1970s that clinicians attempted to categorize muscle pain conditions into distinct syndromes, with specific criteria applying to each.[9–12] In theory, if the patient's condition satisfies the set criteria, a definite diagnosis can be made. The distinction is that the clinician is diagnosing a syndrome, rather than a pathology.

Muscle pain syndromes are generally classified into two distinct categories: fibromyalgia and myofascial pain syndrome, although based on current evidence fibromyalgia is no longer considered a strict "muscle pain" syndrome. To be inclusive, a third category—soft tissue mechanical dysfunction—should be added. Although there are overlapping characteristics of

The author wishes to express gratitude to Christian Gröbli, PT, for his outstanding contributions to the section on myofascial pain; to Mona L. Mendelson, MSW, LCSW-C, for her ongoing support and patience; and to David Simons, MD, for his critical review of this chapter.

these pain syndromes, they represent different neuromusculoskeletal conditions. Soft tissue mechanical dysfunction has a strict mechanical etiology, whereas fibromyalgia and myofascial pain can be caused by mechanical dysfunction or neuro-endocrine or metabolic dysfunction. Examples of soft tissue mechanical dysfunction include partial or full muscle tears or tendinitis. By definition, soft tissue mechanical dysfunction is an acute and local problem usually confined to a particular muscle or tendon. Myofascial pain syndrome is often viewed as a regional pain problem; however, it can be regional or widespread. Myofascial pain syndrome can be acute or chronic in nature. Fibromyalgia is always widespread and chronic. The purpose of this chapter is to explore the etiology, symptomatology, pathophysiology, and medical/therapeutic management of these common pain syndromes, and to discuss the role of the physical therapist and physician in the evaluation and treatment of patients with these conditions.

FIBROMYALGIA

Definition

Fibromyalgia is a disorder of chronic widespread pain, accompanied by tenderness, fa-tigue, sleep disturbance, and psychological distress. Several other syndromes and clinical entities have been linked to fibromyalgia including headaches, irritable bowel syndrome, chronic fatigue syndrome, interstitial cystitis, depression, panic disorder, dyspareunia, endocrine dysfunction involving the hypothalamic–pituitary–adrenal axis, restless leg syndrome, attention deficit hyperactivity disorder, and non-cardiac chest pain.[13-23] Because of its association with so many other syndromes, it has been suggested that fibromyalgia may be part of a broader neuro-endocrine "dysfunctional spectrum syndrome."[24-26]

In North America, fibromyalgia affects 2% of all adults (3.4% of women and 0.5% of men). Seventy to 90% of patients are women. Fibromyalgia is often reported to be a disorder affecting primarily young women, yet it is most common in women ages 50 years and above (Figure 6–1).[15] In a recent study, the prevalence of fibromyalgia under children was 1.2%.[27]

Diagnosis

Following a 1977 publication of Smythe and Moldofsky, a renewed interest in defining criteria for diagnosis and classification of fibrositis emerged, resulting in the 1990 American Col-

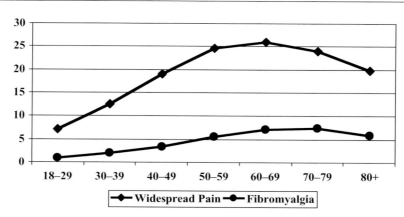

Figure 6–1 Prevalence of widespread pain and fibromyalgia. *Source:* Reprinted with permission from F. Wolfe, K. Ross, et al., The Prevalence and Characteristics of Fibromyalgia in the General Population, *Arthritis & Rheumatism*, No. 38, pp. 19–28, © 1995, American College of Rheumatology, Lippincott Williams & Wilkins.

lege of Rheumatology Criteria for the Classification of Fibromyalgia (ACR criteria).[9,28] The guidelines suggest that a diagnosis of fibromyalgia can be made if a combination of the following criteria is satisfied:

- History of widespread pain (defined as pain in the left side of the body, pain in the right side of the body, pain above the waist, and pain below the waist. In addition, axial pain must be present). Widespread pain must have been present for at least three months.
- Pain in 11 out of 18 anatomically defined tender spots when palpated with approximately 4 kilograms of force. The tender point sites include the following nine paired locations (Figure 6–2):

Occiput:	at the suboccipital muscle insertions
Low cervical:	at the anterior aspects of the intertransverse spaces at C5–C7
Trapezius:	at the midpoint of the upper border
Supraspinatus:	at origins, above the scapula spine near the medial border
Second rib:	at the second costochondral junctions, just lateral to the junctions on upper surfaces
Lateral epicondyle:	2 cm distal to the epicondyles
Gluteal:	in upper outer quadrants of buttocks in anterior fold of muscle
Greater trochanter:	posterior to the trochanteric prominence
Knee:	at the medial fat pad proximal to the joint line

It is noteworthy that the ACR criteria do not include the typical symptoms of sleep disturbance, fatigue, stiffness, and psychological distress. Sleep disturbance, fatigue, and stiffness were found in more than 75% of fibromyalgia

patients; however, the combination of the three symptoms was present in only 56% of patients and lacked the "high sensitivity, specificity and accuracy of the tender point count."[28] The report did suggest considering these typical symptoms; however, they were not essential for categorization purposes.[28] In clinical practice, it appears that many clinicians make the diagnosis of fibromyalgia primarily based on the tender point count in combination with the patient's history.

The ACR criteria have provided researchers with a somewhat homogeneous group of subjects, which has contributed significantly to the publication of more than 1,000 articles on fibromyalgia in peer-reviewed journals during the last decade. It must be emphasized that the criteria are classification criteria established exclusively for clinical and epidemiologic *research* purposes and not for clinical diagnosis, although the criteria suggested that, given their high sensitivity, "they may be useful for diagnosis as well as classification."[28] In 1995, Wolfe and colleagues confirmed that the criteria can be used for clinical diagnosis.[15] As part of the 1992 Second World Congress on Myofascial Pain and Fibromyalgia in Copenhagen, a consensus document on fibromyalgia was defined that emphasized strict adherence to the tender point count in research protocols. According to the Copenhagen declaration, when the ACR criteria are used as "diagnostic" criteria, the diagnosis of fibromyalgia can be made with less than 11 tender points.[29] Other authors, including several contributors to the ACR criteria, also advocate making the clinical diagnosis of fibromyalgia when less then 11 tender points are present, as long as "there are sufficient numbers of fibromyalgia features (e.g., fatigue, sleep disturbance, irritable bowel syndrome, etc.) that are present at a sufficient level of intensity."[30,31] The underlying assumption is that fibromyalgia represents a continuum of distress rather than a discrete syndrome. The number of tender points depicts a more general measure of distress. A high tender point count may indicate more somatic symptoms, more severe fatigue, and low levels of self-care.[32] The specific nature of the

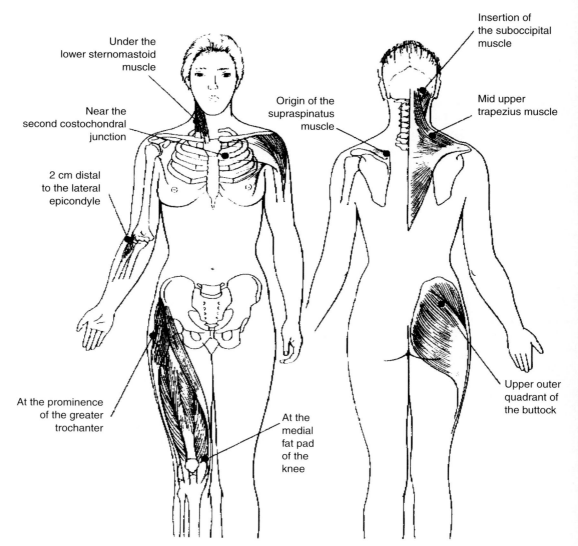

Figure 6–2 Fibromyalgia tender points. *Source:* Reprinted with permission. D.L. Goldenberg. Diagnostic and Therapeutic Challenges of Fibromyalgia, *Hospital Practices* 1989;24(9A):39. © 1989 The McGraw-Hill Companies, Inc. Illustration by Laura Duprey.

fibromyalgia concept, the ACR criteria, and the specificity of the tender points in relationship to fibromyalgia becomes somewhat questionable outside the realm of research, when experts agree that "some loosening of the ACR clas-

sification criteria are necessary for diagnosis in the clinic."[33,34] Perhaps, a focus on tender points is less important than paying attention to the overall psychosocial, behavioral, and organic aspects of individuals with chronic widespread pain.[35] Jacobs and colleagues did not find a correlation between the tender point count and self-reported pain in fibromyalgia.[36]

Another important aspect of the ACR criteria is that the diagnosis of fibromyalgia is made "irrespective of other diagnoses."[28] Therefore, the diagnosis of fibromyalgia is "a diagnosis of inclusion." According to the criteria, a subject meeting the ACR criteria should always be classified as having fibromyalgia. Although this may be satisfactory for classification purposes, it becomes more complicated in clinical diagnosis, especially when there is a treatable condition that also features widespread pain. These conditions include myofascial pain syndrome, myalgia as a complication of cholesterol-lowering medications, hypothyroidism, myoadenylate deaminase deficiency, hypermobility syndrome, or other rheumatic diseases.[37–41] Wolfe maintained that "a person with widespread burns would meet classification criteria for fibromyalgia, but would not be diagnosed as having the syndrome."[42] Although this may seem obvious in the case of a person with widespread burns, it is conceivable that clinicians diagnose fibromyalgia without ruling out other, less obvious pain syndromes.[42] Physicians and physical therapists without adequate training in identifying myofascial trigger points may conclude that there are no other underlying musculoskeletal causes of widespread pain and label a patient incorrectly with fibromyalgia syndrome. An incomplete history may not reveal that a patient started taking cholesterol-lowering medication shortly before the onset of the widespread pain. Others may resort to the diagnosis for difficult-to-categorize patients with significant psychological problems or with any widespread pain.[43] One could argue that in such instances fibromyalgia may be iatrogenically maintained, as appropriate intervention for possible other underlying diagnoses would not be considered.

Because the ACR criteria were not developed for diagnostic purposes, they do not consider the potentially devastating psychological and emotional consequences of "a diagnosis of inclusion" for patients and their families. From a hermeneutic phenomenological perspective, every individual has a strong drive to function in a world of meaning, which can be described as "an individual's transaction with a situation such that the situation constitutes the individual and the individual constitutes the situation."[44,45] In other words, once a person has been given the diagnosis of fibromyalgia, a process may be initiated within that individual that serves to give new meaning to his or her life. Because the current treatment modalities for fibromyalgia have not been able to relieve the symptoms adequately, persons diagnosed with fibromyalgia may actually develop a sense of hopelessness and exhibit iatrogenic illness behavior.[46–49] Specific patient beliefs, including a sense of hopelessness or a belief that one is disabled, are predictive of patient physical and psychological dysfunction and pain behaviors. It is likely that patients with fibromyalgia also adjust to living with the syndrome, frequent pain, loss of hope, and poor expectations, rather than focus on a positive treatment outcome. Hence, they are unnecessarily complicating the rehabilitation process.[50,51] A recent phenomenological study revealed that persons with fibromyalgia appeared to seek constant confirmation of their illness.[52] In some cases, the diagnosis may in fact relieve patients of a sense of responsibility and become a substitute for already troubled life circumstances. In comparing patients diagnosed with fibromyalgia with patients who met the criteria for fibromyalgia but who were not diagnosed as such, it was found that the diagnosed fibromyalgia patients had significantly higher rates of lifetime psychiatric illnesses. The psychiatric diagnoses in these patients were found to be related to "health care seeking behavior" and not to the fibromyalgia. The researchers concluded that "multiple lifetime psychiatric diagnoses may contribute to the decision to seek medical care for fibromyalgia in tertiary care settings."[53] There is some controversy whether making a diagnosis of fibromyalgia reduces patients' utilization of medical resources, or actually facilitates a dependency on the medical system.[54,55] McBeth and colleagues established that a high tender point count was associated with increased medical care usage in addition to an increased number of physical symptoms.[32]

Therefore, it may not always be in the best interest of a patient to be diagnosed with fibromyalgia, especially in the presence of other musculoskeletal pain diagnoses for which there are potential solutions.[42] For classification purposes, a "diagnosis of inclusion" may be appropriate, even though it may still influence the outcome and conclusions drawn from such research. For example, a large cohort study of the intermediate and long-term outcomes of fibromyalgia in patients seen at least once in specialty rheumatology clinics concluded that the prognosis of fibromyalgia was very poor. Although the study did not examine the results of treatment at these centers, patients with fibromyalgia continued to demonstrate significant clinical and functional abnormalities.[48] Again, it is possible that clinicians did not consider the other diagnoses of widespread pain as the ACR criteria were applied. Perhaps, the patients were not evaluated for the presence of myofascial pain syndrome, hypothyroidism, or other differential diagnoses and did not receive the most appropriate treatment. For clinical purposes, the diagnosis of fibromyalgia should be made as a "diagnosis of exclusion." It is not sufficient to diagnose fibromyalgia without considering all differential diagnoses. This does not necessarily reflect a clinician's attitude that fibromyalgia does not exist, but assures patients of the most appropriate treatment. The medical and therapeutic management should focus primarily on resolving the other diagnoses and not resort to just "teaching patients how to manage their fibromyalgia." This was illustrated by Poduri and Gibson, who described a case of a patient who met the ACR criteria and who was subsequently diagnosed with fibromyalgia, but who in fact suffered from drug-related lupus and required immediate treatment accordingly.[56]

To overcome some of the clinical limitations of the ACR criteria, Müller and Müller developed specific *diagnostic* criteria for fibromyalgia.[57] When 11 of 18 tender points must be painful with 4 kg/cm² for fulfilling the ACR criteria, Müller and Müller required 12 of 24 points to be tender when pressed with a force of 2 kg/cm² using dolorimetry. They found that their method had a significantly higher specificity with a similar sensitivity, when compared with the ACR proposed method.[58] In summary, the Müller and Müller criteria require:

- Spontaneous pain in muscles, in the course of tendons or at tendon insertions in at least three body regions of the trunk and extremities for at least three months.
- Decreased pain threshold with a visible pain response following digital pressure of 2 kg/cm² of 12 out of 24 tender points.

In addition, Müller and Müller suggested the fulfillment of secondary criteria, including autonomic symptoms, functional limitations, and psychopathological findings. The finding of three autonomic and three functional symptoms would further support the diagnosis of fibromyalgia. Autonomic symptoms may include cold hands or feet, dry mouth, hyperhidrosis, excessive dermographia, orthostatic symptoms, respiratory arrhythmia, and tremor. Functional limitations may include sleep disturbances, gastrointestinal problems, cardiac problems, paresthesia, and dysuria/dysmenorrhea. Müller and Müller agreed that the diagnosis of fibromyalgia should be made only as a "diagnosis of exclusion."[57–59]

In spite of extensive research efforts, there are no objective laboratory studies that confirm the diagnosis of fibromyalgia. The ACR criteria were developed solely based on consensus, a process not uncommon in medicine, that has been applied to defining several other clinical entities, including migraine and depression.[60] The reliance on a definition by consensus and the lack of a well-defined concept of pathophysiology have resulted in critical opposition to the fibromyalgia construct. The ACR criteria have been criticized as being arbitrary and at risk for circular reasoning and tautology.[61–63] It appears that the same criticism would apply to the Müller and Müller criteria.

Clinical Characteristics

Clinically, the patient typically has complaints of diffuse and widespread pain that is, however, not confined to tender points.[64,65] Almost all patients report significant sleep disturbances and report arising in the morning feeling unrefreshed and physically fatigued. Depression, anxiety, and somatization are frequently reported in studies of fibromyalgia patients. Other clinical presentations may include hypersensitivity to cold or heat, frequent bouts of abdominal pain, constipation and diarrhea, recurrent frontal-occipital headaches, and sensations of numbness or swelling in the hands and feet. The patient usually describes chronic pain, headaches, and fatigue experienced for many years. The result of x-rays, computed tomography (CT) scans, magnetic resonance imaging (MRI), electromyography, and blood studies are normal, and should not be ordered unless other clinical symptoms would indicate such.

Tenderness

An essential feature of the ACR criteria is a total tender point count of at least 11 out of 18 anatomically defined points, when these points are subjected to 4 kg of digital pressure. Semantically, it is interesting that the ACR criteria require 11 out of 18 anatomically defined tender points, rather than 11 tender points out of 18 anatomically defined points. Fibromyalgia tender points need to be distinguished from myofascial trigger points, which are the main characteristic of the myofascial pain syndrome. In fibromyalgia, hyperalgesia is not limited to the tender point, but expresses a more generalized and widespread pain problem.[64–66] The fibromyalgia tender points do not have an established physiological or histopathological basis like myofascial trigger points. Trigger points are actual contraction knots in muscles that refer pain to a more distant region. Patients with fibromyalgia may also have myofascial pain syndrome and myofascial trigger points, yet "fibromyalgia trigger points" do not exist.[67,68] There is no evidence that myofascial pain can evolve into fibromyalgia.

Although the validity and inter-observer and intra-observer reliability of the tender point count have been established in several studies, Fischer commented that physicians were not able to provide exactly 4 kg/cm^2 or reproduce the same level of pressure on repeated attempts.[69–72] To overcome possible inaccuracies, the use of pressure algometry is recommended. Pressure algometry is a standardized method for quantification of tenderness and is expressed by the so-called pressure pain threshold, or the minimum pressure that induces pain or discomfort.[73] It is not only useful for diagnostic purposes, but also provides a means to evaluate immediate and long-term treatment outcomes.[68,73,74] Zohn and Clauw explored the utility of skin rolling as a clinical test for fibromyalgia and found that skin rolling approached the reliability of the tender point count.[75] Skin rolling does not depend on a verbal response of the patient and may be more objective than a tender point count; however, at this point, further studies are needed to establish the nature of skin rolling and its applicability to the diagnosis of fibromyalgia.

Persons with fibromyalgia have altered nociception and hypervigilance.[76–79] Vecchiet and colleagues measured the sensitivity to electrical stimulation of the skin, the subcutis, and the muscle and observed that hyperalgesia in all three tissues was present not only over fibromyalgia tender points, but also in nonpainful regions.[66] In another study, the pressure-induced pain sensitivity in fibromyalgia patients was more pronounced deep to the skin and not restricted to muscle tissue. The altered sensitivity was not dependent on increased skin sensibility.[65,80] Gibson and colleagues demonstrated that persons with fibromyalgia exhibited a significant reduction in heat pain threshold as well, although this was not confirmed by others.[81,82] In spite of these findings, there is no convincing evidence that the peripheral tissues in persons with fibromyalgia are abnormal.[83–85] Graven-Nielsen and colleagues concluded that the hyperalgesia observed following painful stimuli of a pain-free muscle in fibromyalgia patients indicates the involvement of central hyperexcitability.[86] Patients with fibromyalgia had a lower resting state level of regional cerebral blood flow in the thalamus and caudate nucleus, which also suggests that central sensitization is the final common pathway for the development of abnormal pain perception.[87,88]

Sleep Disturbances

Many persons with fibromyalgia report waking up unrefreshed and fatigued, but this symptom is not universal. In some persons, fatigue may be debilitating, whereas in others it is absent or has been accepted because of its chronic nature.[89] Fatigue may be the result of disturbed sleep, which in itself is a factor positively associated with self-reported work disability.[48,90] In general, sleep is characterized by alternating cycles of rapid eye movement (REM) sleep and non-REM sleep. Non-REM sleep is divided into four stages with increasing percentages of low frequency brain waves referred to as delta waves. Stages 3 and 4 feature predominantly delta waves and are referred to as "deep sleep" or "slow wave sleep." It is during these stages that restorative sleep occurs.[91,92] In 1975, Moldofsky and colleagues reported that fibromyalgia patients have an abnormal sleep pattern characterized by the so-called alpha-delta anomaly, an intrusion of alpha waves during slow wave sleep.[93] Although several studies have confirmed Moldosky's observations, others failed to duplicate their findings.[94–98] The alpha-delta sleep anomaly was found in only 36% of fibromyalgia patients and was not specific for fibromyalgia.[99] It has been described in persons with acquired immune deficiency syndrome (AIDS), osteoarthritis, rheumatoid arthritis, and even in healthy subjects.[100–103] Scudds and colleagues did not find any difference for sleep quality between persons with fibromyalgia and persons with myofascial pain syndrome.[104] Lue questioned the sensitivity and specificity of alpha electroencephalography.[105] At this time, there is insufficient evidence that disturbed sleep patterns are specific for persons with fibromyalgia; any chronic pain state appears to have a negative effect on a person's sleep pattern.

Psychosocial Factors

Most studies demonstrated that persons with fibromyalgia have more emotional and psychological problems than persons with other chronic pain syndromes and normal control subjects, which led Hudson and Pope to conclude that fibromyalgia is an "affective spectrum disorder."[106] According to Yunus, the use of this term is inappropriate because nearly all reports on psychological dysfunction are based on patients seen in specialty rheumatology clinics and may over-represent psychological problems based on referral bias.[107] Walter and colleagues also concluded that affective distress is not unique to fibromyalgia, but primarily the result of higher levels of pain severity.[108] In an older study, no significant differences were found between persons with fibromyalgia and control subjects in a general medicine clinic.[109]

Although there is no evidence that there is a "fibromyalgia personality," a few studies identified a "pain prone personality" in some patients with fibromyalgia.[110,111] Persons with a pain prone personality are typically high achievers, who lack assertiveness and the ability to perceive and express unpleasant emotions. The development of a pain prone personality may be related to posttraumatic stress disorder, repetitive trauma, and adverse childhood experiences.[111,112] Twenty-one percent of patients with fibromyalgia were found to suffer from posttraumatic stress disorder versus none of the control group, while several other studies have linked alcoholism in families and sexual and physical abuse to fibromyalgia.[113–117] Adverse childhood experiences were positively correlated with a high tender point count.[32] A pain prone personality is not specific to fibromyalgia and is seen among a broad spectrum of psychosomatic and adjustment disorders.[111]

Patients with fibromyalgia appear to have higher rates of lifetime and current depression, notwithstanding a few studies that did not find any evidence of increased depression.[21,118–121] As with most symptoms of fibromyalgia, it is not clear how the symptoms are related to the fibromyalgia. Do patients with fibromyalgia get depressed because of pain, or can depression cause or significantly contribute to fibromyalgia due to increased pain sensitivity? Or are both disorders the result of a common underlying abnormality? Based on recent studies and theories, depression and fibromyalgia are most likely the result of a common underlying abnormality, perhaps an insufficient catecholaminergic or serotonergic neurotransmission or hyperactivity of corticotropin-releasing hormone.[18,121,122] It is likely that having a diagnosis of fibromyalgia combined with constant pain, poor expectations regarding recovery, and a sense of hopelessness, however, may also become perpetuating factors for depressive mood disorders. Fassbender and colleagues observed that patients with fibromyalgia had significantly more tender points than patients with depression.[123] Patients with fibromyalgia demonstrated significantly higher lifetime prevalence rates of mood, anxiety, and somatization disorders than patients with rheumatoid arthritis.[115,124–126] Wolfe and colleagues found that persons with fibromyalgia are more than four times as likely to be divorced compared to the general population without fibromyalgia.[15]

Several authors have suggested that fibromyalgia is "just another somatization disorder."[127] Hellström and colleagues pointed out that "to put a label on suffering gives it meaning."[52] Having a diagnosis of fibromyalgia may provide a means to avoid dealing with psychosocial issues or justify why patients are not really "responsible for their perceived inability to comply with the demands they themselves and others would place upon them."[52] Ford also considered fibromyalgia a form of somatization and a "fashionable diagnosis" and agreed that somatization could serve as a rationalization for psychosocial problems or as a coping mechanism.[128] Fibromyalgia can become "a way of life," or as Hadler stated, "if you have to prove you are ill, you can't get better."[46,128] This becomes particularly difficult in determining whether persons with fibromyalgia should be awarded disability or injury compensation. A recent epidemiological study concluded that having been told that one had fibromyalgia became one of the predictors of self-reported work disability, which illustrates the influence of cognitive beliefs on somatic perception.[90] Similarly, Haynes and colleagues established that patients who did not know that they were hypertensive had a threefold increase in work absenteeism after being told the diagnosis.[129] If having been diagnosed with fibromyalgia is a factor in self-reported work disability, and if the symptomatology provides an extenuation of an already difficult life, should persons with fibromyalgia receive disability benefits? Although the majority of persons with fibromyalgia report being able to work, as many as 25% have received some form of compensation.[43,90,130]

Barsky and Borus included fibromyalgia in their description of "functional somatic syndromes," a group of syndromes characterized more by symptoms, suffering, and disability than by objective findings. According to the authors, these syndromes include multiple chemical sensitivity, the sick building syndrome, repetition stress injury, the side effects of silicone breast implants, the Gulf War syndrome, chronic whiplash, the chronic fatigue syndrome, and the irritable bowel syndrome.[131] Functional somatic syndromes have certain characteristics in common. Persons suffering from any of these syndromes often attribute common somatic symptoms to the illness. Common symptoms are amplified and become the main focus of attention. They are convinced that they have a serious illness that is likely to worsen. The more a patient is convinced of having a serious illness, the stronger the tendency to search for a confirmative diagnosis. Confirmation of illness seemed to be important for persons with fibromyalgia.[52] Wolfe confirmed that persons with fibromyalgia reported more medical conditions and perceived more significance to these conditions than persons with rheumatoid arthritis or osteoarthritis.[55] Many patients with functional somatic syndromes assume the "sick role," which is further exacerbated by litigation, disability compensation, and society's portrayal of the condition as "catastrophic and disabling." Barsky and Borus outlined several other factors relevant for the discussion of fibromyalgia. Health care institutions, medical providers, and advocacy groups have developed professional and financial interests in the diagnosis, as evidenced by the increasing number of fibromyalgia clinics, Internet Web sites devoted to fibromyalgia, and the multiple support groups, which will reinforce the belief that there is no effective treatment (Table 6–1).[132,133]

Many patients with fibromyalgia have adopted other diagnoses and feel that they also have chronic fatigue syndrome or irritable bowel syndrome, a process sometimes referred to as "pathoplasticity," realizing that these additional syndromes may have etiologic similarities to fibromyalgia.[134,135] The diagnosis given to a patient may in fact depend on the specialty of the physician. A rheumatologist may diagnose fibromyalgia, an internist may identify chronic fatigue syndrome, while a gastroenterologist may consider irritable bowel syndrome. In spite of these controversies, patients with fibromyalgia or chronic widespread pain will continue to seek medical help irrespective of physicians' belief systems.[136]

Lack of Exercise

Lack of exercise is another relevant factor in the clinical history and presentation of fibromyalgia. Most persons with fibromyalgia exercise little and assume that exercise will worsen their condition. Persons with fibromyalgia tend to be deconditioned, which may account for some of the apparent abnormalities reported in oxygen consumption and accumulation of metabolites.[137,138] When compared to equally fit healthy subjects, however, persons with fibromyalgia were found to have normal oxygen consumption and normal accumulation of metabolites during exercise.[139–141] Other studies demonstrated that there was no increased structural damage with exercise when compared with healthy individuals.[142–144] Although the number of subjects was limited, a few studies suggested that persons with fibromyalgia may have a hyporesponsiveness of the sympathetic nervous system and hypothalamus–pituitary–adrenal axis during exercise.[145,146]

Pathogenesis

One of the difficulties of diagnosing and treating patients with fibromyalgia is the absence of findings in the laboratory and radiologic workup. Much research has been conducted to identify histological and physiological characteristics of fibromyalgia to determine possible etiologies and effective treatment remedies. Fibromyalgia is a complex, multi-factorial disorder that has been associated with musculoskeletal and neurochemical abnormalities, yet most of these abnormalities are not specific for fibromyalgia. None of the findings have resulted in fibromyalgia-specific laboratory studies or objective diagnostic criteria. Initial studies attempted to identify musculoskeletal abnormalities and signs of inflammation. Altered muscle metabolism, decreased circulation, and structural damage to muscles have been suggested to explain the widespread muscle pain in patients with fibromyalgia. More recent research has focused on the role of neurotransmitters, the hypothalamus–pituitary–adrenal axis, and various hormones. A brief review of pertinent research follows.

Musculoskeletal Abnormalities

Several studies identified "rubber bands" in single muscle fibers, "moth-eaten" and "ragged

Table 6–1 Number of Web Sites Found on www.altavista.com (January 10, 2000)

Search Word	Number of Sites
Heart disease	249,547
Arthritis	428,885
Cancer	2,181,318
AIDS	2,321,925
Fibromyalgia chat	14,373,294
Fibromyalgia	87,726,785

red" fibers, a reduced content of high energy phosphates, and a higher rate of phosphodiester resonance, which were thought to be related to damage to the sarcolemma, an abnormal occurrence of elastic fibers, an energy deficiency state, or local muscle hypoxia.[143,147-153] "Motheaten" fibers are indicative of a change in the distribution of mitochondria or the sarcotubular system; "ragged red" fibers reflect an accumulation of mitochondria.[154] Sprott and colleagues identified decreased levels of collagen cross-links in persons with fibromyalgia, suggestive of altered collagen metabolism, which may contribute to remodeling of the extracellular matrix. They hypothesized that these changes may contribute to the lowered pain threshold at tender points.[155] Others did not find any significant differences between fibromyalgia and normal muscles.[83,85,156,157] When fibromyalgia patients were matched with equally fit, healthy control subjects, no differences were found in lactate and potassium levels, oxygen uptake, and P^{31} magnetic resonance spectroscopy, suggesting that patients with fibromyalgia do not have abnormal muscle metabolism.[139-142,158,159] There is also no evidence of any structural damage to muscles of persons with fibromyalgia.[138,142] Magnetic resonance imaging did not reveal any abnormalities of the skeletal muscles of persons with fibromyalgia.[160] The structural and functional abnormalities noted in earlier studies appear to be the result of muscle deconditioning and are not specific for fibromyalgia.[161] Because of the lack of specific peripheral and histological findings, the focus of fibromyalgia research has shifted toward investigations of the central nervous system and the endocrine system. To understand the neurophysiological mechanisms responsible for pain syndromes, it is critical to integrate knowledge from the pain sciences into clinical practice.[162]

Neurochemical Abnormalities

Substance P. Several studies have identified substance P levels to be up to three times higher in the cerebrospinal fluid of persons with fibromyalgia compared with healthy control subjects.[163-165] Patients with painful hip osteoarthritis were found to have 1.5 to 2 .0 times normal levels of substance P, whereas patients with neuropathies, including diabetic neuropathy, had either below normal or 1.5 times normal levels.[166-168] Substance P is a neuropeptide involved in several aspects of the process of nociception. It is released in the dorsal horn of the spinal cord in laminae I, II and V ($A\delta$ fibers) and laminae I and II (C fibers) by activated $A\delta$ and C fiber afferent neurons. This seems to suggest that there is a peripheral origin of the nociceptive stimuli; however, at this point, there is no evidence to support a peripheral mechanism in fibromyalgia.[122,169] The large diameter sensory fibers ($A\beta$) are non-nociceptive and terminate in laminae III and IV. They do not contain neuropeptides, but release glutamate as their neurotransmitter.[170] Dorsal horn neurons are divided into high-threshold mechanosensitive neurons, low-threshold mechanosensitive neurons, wide-dynamic-range neurons, and interneurons. All neurons can be directly sensitized or develop new synaptic contacts with other neurons. A long-term increase in the excitability of wide-dynamic-range neurons following a noxious event may contribute to the development of chronic pain disorders.[171] Under normal circumstances, high-threshold mechanosensitive neurons are synaptically connected with $A\delta$ and C fibers. Therefore, they respond to noxious stimuli, whereas low-threshold mechanosensitive neurons do not mediate pain. Afferent barrage from joints, skin, viscera, and muscles can unmask previously ineffective, or "sleeping" synapses within the dorsal horn by the release of substance P, calcitonin-gene related peptides, and glutamate from the primary afferent neuron into the dorsal horn via *N*-methyl-D-aspartate and neurokinin-1 receptors. There is some evidence that $A\beta$ fibers sprout dorsally from laminae III and IV into laminae I and II following peripheral injury, resulting in new synapses with nociceptive neurons. Low-threshold ($A\beta$) afferent input would then be interpreted as noxious.[170] Because substance P can lower the threshold of synaptic excitability, there may be an increase in the number of mechanosensitive receptive fields, making fibromyalgia a syndrome of central sensitization.[172–174]

The pain in fibromyalgia may be related to the action of substance P on neurokinin-1 effector receptors that promote nociception. This does not explain the widespread nature of the pain, however, because the spread of excitation in the spinal cord is fairly limited.[175] Giovengo and colleagues reported the findings of elevated levels of nerve growth factor in the cerebrospinal fluid of persons with fibromyalgia. Nerve growth factor is thought to facilitate the growth of substance P containing neurons and increase the excitability of dorsal horn cells by afferent muscle input.[176,177] The nociceptive activity of substance P is counteracted by serotonin, which can inhibit spinal nociceptors via the descending antinociceptive pathways.

Serotonin. Serotonin (5-hydroxytryptamine) is a neurotransmitter involved in the organization of mood, emotion, motor functions, sleep, neuroendocrine rhythms, cognition, and pain perception.[178] It is one of the neurotransmitters responsible for regulation of the function of the hypothalamic–pituitary–adrenal axis. Serotonin can influence the release of corticotropin-releasing hormone from the hypothalamus. It also triggers the release of adrenocorticotropin hormone from the anterior pituitary and exerts a direct influence on the corticosteroid production from adrenocorticol cells. Serotonin increases the production of cyclic adenosine monophosphate.[179] It is not known whether serotonin deficiencies will result in the perturbations of the hypothalamic–pituitary–adrenal axis seen in persons with fibromyalgia.[180–182] Multiple serotonin receptor sites have been identified in the gastrointestinal tract, which may be relevant given the relative common occurrence of functional bowel disorders in person with fibromyalgia.[183]

Moldofsky and colleagues were the first to link serotonin levels with fibromyalgia, because of its role in the initiation and perpetuation of slow wave sleep and the regulation of pain perception through activity in the thalamus.[184,185]

They reported a correlation between pain in fibromyalgia and the plasma concentration of the essential amino acid tryptophan. Tryptophan is the metabolic precursor to serotonin that is extracted from dietary protein in the intestines. Tryptophan is oxidatively decarboxylated to serotonin by neurons in the brain stem raphne nucleus, which is then released in the brain and spinal cord. In experiments with rats, serotonin enhanced the synthesis of substance P in the brain, while it inhibited the release of substance P in the spinal cord. It is likely that persons with fibromyalgia have low brain tissue levels of both serotonin and substance P, and low spinal cord levels of serotonin and high spinal cord levels of substance P.[186] Although spinal cord levels of serotonin have not been reported in cerebrospinal fluid of persons with fibromyalgia, the concentrations of its immediate precursor 5-hydroxy-tryptophan and its metabolic product 5-hydroxy-indole acetic acid were found to be lower when compared to normal control subjects.[187–189] Lower serum levels of both tryptophan and serotonin have been reported, possibly related to the diversion of tryptophan into kynurenine instead of serotonin and to low levels of platelet serotonin.[188,190–192] The range of serum levels of serotonin in fibromyalgia patients appeared to be large and may not be consistently correlated with fibromyalgia symptoms, including depression, tender points, and dolorimetry.[193] Ernberg and colleagues found higher levels of serotonin in the superficial masseter muscles of patients with fibromyalgia compared with healthy control subjects. The higher levels appeared to originate in the blood supply, but could also be peripherally released.[194] Klein and colleagues reported the presence of antibodies against serotonin, phospholipids, and gangliosides, which are part of the serotonin receptor, in patients with fibromyalgia.[14,195–197] Antibodies against serotonin were also reported in persons with panic disorder, a condition sometimes associated with fibromyalgia.[198] The inhibition of spinal nociceptors via descending pathways is accomplished primarily via serotonergic and noradrenergic neurons.[174,199] Perhaps the widespread pain in fibromyalgia is the result of a dysfunction of the descending antinociceptive system or of an overactivity of the descending pathways that facilitate nociception.[200,201]

Hormonal Abnormalities. Because the onset of fibromyalgia is often reported to coincide with physical or emotional stress, it is not surprising that several researchers have focused on possible disturbances of the stress response systems, including the hypothalamic–pituitary–adrenal axis and the sympathetic nervous system

(Figure 6–3).[22,122,202–205] Fibromyalgia can be considered a "stress-related syndrome."[203] The hypothalamic–pituitary–adrenal axis is the main physiologic response system to stress.

Regulation of the hypothalamic–pituitary–adrenal axis occurs primarily through modulation of corticotropin-releasing hormone, an amino acid peptide that stimulates the secretion of adrenocorticotropic hormone and other hormones. Adrenocorticotropic hormone is an anterior pituitary peptide that stimulates the secretion of glucocorticoids and other steroids from

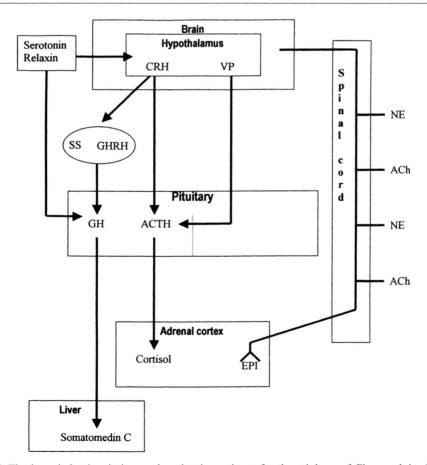

Figure 6–3 The hypothalamic–pituitary–adrenal axis pertinent for the etiology of fibromyalgia. *Note:* CRH, corticotropin-releasing hormone; VP, vasopressin; SS, somatostatin; GHRH, growth hormone releasing hormone; GH, growth hormone; ACTH, adrenocorticotropic hormone; NE, norepinephrine; ACh, acetylcholine; EPI, epinephrine.

the adrenal cortex. Cortisol is the main form of glucocorticoids released in humans.[180,206] Corticotropin-releasing hormone stimulates adrenocorticotropic hormone in a diurnal rhythm with a peak before awakening and a decline as the day progresses. The diurnal rhythm of adrenocorticotropic hormone is reflected in the diurnal secretion of cortisol.[207] When a stressor is perceived by the brain, corticotropin-releasing hormone is released.[180,182,206] The activity of corticotropin-releasing hormone neurons appears to determine several of the symptoms of fibromyalgia.[122,205] Persons with fibromyalgia displayed a hyperreactive adrenocorticotropic hormone release and a blunted cortisol release in response to exogenous corticotropin-releasing hormone and to endogenous activation by insulin-induced hypoglycemia.[204,208] The release of adrenocorticotropic hormone by corticotropin-releasing hormone is augmented by arginine vasopressin, another hypothalamic peptide. Based on studies of rats, arginine vasopressin may be instrumental in maintaining the activation of the hypothalamic–pituitary–adrenal axis during chronic stress.[19,209,210] Different stressors cause different patterns of release of the hypothalamic hormones. Riedel and colleagues observed elevated basal levels of adrenocorticotropic hormone and cortisol in fibromyalgia patients.[205] Crofford and colleagues and McCain and Tilbe found normal morning levels of cortisol, but elevated evening levels, resulting in a loss of the normal diurnal cortisol fluctuation.[202,203] Reduced 24-hour urinary free cortisol levels were found as compared with normal subjects and persons with rheumatoid arthritis or low back pain, especially in persons with longstanding fibromyalgia.[202,203,208,211] Crofford and Demitrack speculated that the apparent discrepancy between elevated evening levels of cortisol and reduced 24-hour levels may be attributed to a reduction of the normal frequency of cortisol release.[181] In contrast with these findings, Adler and colleagues found normal 24-hour urinary free cortisol levels and normal diurnal patterns of adrenocorticotropic hormone and cortisol.[22] They found a 30% reduction in adrenocorticotropic hormone and epinephrine responses to hypoglycemia, contrasting the findings by Griep and colleagues of an exaggerated adrenocorticotropic hormone response.[22,204] Nevertheless, they agreed that fibromyalgia may be primarily characterized by an impaired hypothalamic–pituitary–adrenal axis.[22]

Another aspect of the hypothalamic–pituitary–adrenal axis was recently investigated by Dessein and colleagues, who looked at the levels of dehydroepiandrosterone sulphate, testosterone, cortisol, serotonin, and insulin-like growth factor-1 (somatomedin C) and their correlation with health status in persons with fibromyalgia.[212] Dehydroepiandrosterone sulphate is the metabolic precursor to estrogen, which was recently shown to be involved in the regulation of enkephalin levels in the superficial dorsal horn, thereby changing the response to nociceptive stimuli.[213] During pregnancy, dehydroepiandrosterone sulphate is involved in the placental production of estradiol.[214] Dehydroepiandrosterone sulphate levels are a good indicator of adrenocortical function and probably more sensitive than cortisol levels.[215] Under stress, the secretion of dehydroepiandrosterone sulphate is diminished. With aging, there is a suppression of dehydroepiandrosterone sulphate secretion, but not of corticosteroid production.[216] Lower habitual physical activity was related to lower levels of circulating dehydroepiandrosterone sulphate and insulin-like growth factor-1 independently of age and anthropometric measures. In healthy elderly women, lower maximal aerobic capacity was associated with lower dehydroepiandrosterone sulphate concentrations.[217] There is also a positive correlation between hours of sleep and serum dehydroepiandrosterone sulphate levels.[218] Dessein and colleagues found that the levels of dehydroepiandrosterone sulphate and testosterone were significantly reduced in women with fibromyalgia. They speculated that the androgens may protect against fibromyalgia. There was a positive correlation between dehydroepiandrosterone sulphate levels and pain, which disappeared after adjusting for increased weight. Only 14% of the subjects were normal weight in this study and there was an association between a high body-mass index and decreased dehydroepiandrosterone sulphate levels, which contradicted the findings by Maccario and colleagues in healthy adults.[212,219] In Maccario's study, the dehydroepiandrosterone-sulphate levels were positively and independently associated with 24-hour urinary cortisol and insulin-like growth factor-1 levels.[219] Dessein and colleagues did not find any significant relationship between the levels of cortisol, serotonin, and insulin-like growth factor-1 and health status as measured by the Fibromyalgia Impact Questionnaire.[212]

Several studies have demonstrated that persons with fibromyalgia may have low levels of growth hormone (somatotropin) and insulin-like growth factor-1.[205,220–224] Growth hormone is an amino acid polypeptide hormone synthesized and secreted by the anterior pituitary. Its primary function is to promote linear growth. Growth hormone stimulates the release of somatomedin C in the liver, which is required for the maintenance of normal muscle homeostasis.[122] Approximately 70% of growth hormone is secreted during slow-wave sleep and the amount of secreted growth hormone correlates with the amount of slow-wave sleep.[225] It was postulated that the poor sleep patterns of persons with fibromyalgia could disrupt the nocturnal secretion of growth hormone.[223] The secretion of growth hormone is under bidirectional control of the hypothalamus, which contains both growth hormone releasing hormone as well as a growth hormone inhibiting hormone, known as somatostatin.[180,182] It is not known whether there is a decrease in growth hormone releasing hormone, or an increase in somatostatin. The somatostatin secretion is promoted by corticotropin-releasing hormone and thyroid hormones, which is another reason to include thyroid dysfunction in the differential diagnosis of fibromyalgia.[206,226–228] Leal-Cerro and colleagues concluded that the decrease in growth hormone secretion was due to hypothalamic dysfunction.[224] Nørregaard and colleagues did not find any differences in somatomedin C levels among persons with fibro-

myalgia compared to healthy, but sedentary control subjects.[229] They suggested that perhaps the difference in findings was due to selection procedures, as it is known that physically active individuals have significantly higher somatomedin C levels than sedentary subjects.[229,230]

An intriguing hypothesis regarding the etiology of fibromyalgia was postulated by Yue.[231] Notwithstanding observations by Ostensen and colleagues describing worsening of symptoms during pregnancy with the last trimester experienced as the worst period, Yue noted that pregnant patients with fibromyalgia often experience a remission of their symptoms during pregnancy with a return of symptoms within one or two months following delivery.[231,232] In addition, Yue found that many patients with fibromyalgia responded positively to injections with botulinum toxin. These findings made Yue search for any agent or hormone that would have an effect on the collagen of connective tissues, which resulted in the hypothesis that the pathogenesis of fibromyalgia is related to a systemic deficit of relaxin, or an inability of the body to utilize relaxin.[231] He speculated that the increased use of birth control pills at a younger age may lead to relaxin deficiencies. A fast onset of fibromyalgia appeared to occur in women following oophorectomies or hysterectomies. In males, low levels of relaxin appeared to be related to low levels of testosterone.

Relaxin is a polypeptide hormone related to insulin and insulin-like growth factors. It is secreted in females in the corpus luteum, decidua, and placenta and in males in the prostate, from which the hormone is secreted mainly in seminal plasma. Relaxin is best known for its role during pregnancy and is known to promote lengthening and softening of pelvic ligaments to facilitate the birth process. Relaxin does not only effect connective tissue extensibility, but plays a role in many other biological processes.[233] It is involved in the inhibition of uterine contractile activity and it stimulates the growth of the mammary gland. In males, relaxin is thought to promote motility of spermatozoa.[234] Relaxin has a strong vasodilatory effect and it promotes the generation of nitric oxide, which also appears to play an important role in muscle pain.[235-237] Of particular interest is that in experiments with rats, relaxin binding sites have been identified in several regions of the brain that are involved in the control of systemic blood pressure and the secretion of hypothalamic hormones.[238,239] Relaxin stimulates the release of oxytocin and vasopressin, which, as discussed above, augments the release of adrenocorticotropic hormone by corticotropin-releasing hormone.[19,240-242] Relaxin was also found to promote the secretion of prolactin and growth hormone.[243,244] Yue speculated that daily administration of relaxin in persons with fibromyalgia may alleviate many of the symptoms.[231] At this point, Yue's speculations are not supported by independent research; however, given the broad spectrum of relaxin this hypothesis deserves further attention.

Management of Fibromyalgia

Given the complexity of fibromyalgia, there is usually no single discipline or treatment remedy that can offer optimal solutions, although Granges and colleagues reported a remission rate of 24% after two years using minimal intervention in a community rheumatology practice.[245,246] As Turk and Okifuji have outlined, the assessment of patients with chronic pain requires attention to relevant psychosocial, behavioral, and organic factors and an integrated interdisciplinary treatment strategy.[35] The available data suggest that the integration of medicine, psychology, and physical therapy offers the best possible treatment outcome.[246] Certain aspects of the treatment can be done in group format, whereas others require individual interventions.[247] One study did not support adding cognitive group treatment to group education.[248,249] It is important that patients are part of the interdisciplinary team and develop clear perceptions about their participatory role as functional members of the health care team.[250] The interdisciplinary approach can lead to significant improvements in pain severity, life interference, sense of control, affective distress, depression, perceived

physical impairment, fatigue, and anxiety.[251] The management approach must be personalized and address the major determinants relevant for the individual patient.[26] In one study, patients with fibromyalgia were classified in one of three groups based on their responses to the Multidimensional Pain Inventory. The "dysfunctional" group was characterized by poor coping and high levels of pain. The "interpersonally distressed" group was characterized by interpersonal problems. The "adaptive copers" demonstrated low levels of affective distress and disability.[252] Following interdisciplinary intervention, statistically significant reductions were observed in pain, affective distress, perceived disability, and perceived interference of pain for the "dysfunctional" group, but not for the "interpersonally distressed" group. The "adaptive copers" did not change that much, possibly because of low pretreatment levels of distress.[252]

All clinicians must recognize the multi-complexity of dealing with persons with chronic pain. They must be cognizant of the differences between acute and chronic pain and appreciate the common lifestyle changes that patients with chronic pain often make. It is imperative to be aware of the "5 Ds" of chronic pain: dramatization, drug misuse, dysfunction, dependency, and disability. The concepts of learned helplessness and hopelessness must be understood to avoid misinterpreting patients' verbal and nonverbal communications. Many chronic pain patients prefer to view their pain condition as a medical problem, externalizing their responsibility for their pain and their life situation. Clinicians working with persons with fibromyalgia must be comfortable with different learning styles and the role of family, work, and the greater community on the patient, and be willing to work with the patient within these environments.[253] Some familiarity with systems theory may be valuable for physicians and physical therapists in working with persons with fibromyalgia.

Systems theory is the most popular theory used in contemporary social work practice that focuses on the interactions and transactions between patients and their environments. It includes the marital relationship, the family and society, as well as functional and structural aspects.[254] Although physical therapists should not engage in offering systems intervention, a system-centered orientation can play an essential role in physical therapy, especially in understanding the broader context in which patients function. Insight into the patients' belief systems and expectations or lack thereof is essential. There is no doubt that assuring patients that their pain is taken seriously, and that their intentions are not questioned by their pain specialists, is critical from the first encounter and throughout the treatment process. Patients need to become active participants in deciding the optimal treatment algorithm.[253]

Bennett emphasized the need to improve self-efficacy by down-regulating the positive feedback loop that exists when stress of chronic pain levels results in physiologic arousal with secondary symptoms.[246] Self-efficacy in exercising control over cognitive stressors can actually activate endogenous opioid systems. The stronger the perceived self-efficacy to withstand pain, the longer subjects endured increasing levels of pain stimulation.[255,256] Bandura described four techniques for altering patients' perception of self-efficacy, including social persuasion, mastery experiences, modeling, and physiologic feedback. Through social persuasion, health care providers and significant others attempt to convince patients that they can be more functional than they perceive. By performing activities that previously were thought to be impossible because of pain or other dilemmas, patients can master new experiences. Exposing persons with fibromyalgia to others with fibromyalgia who have succeeded in changing their lives and becoming more functional can provide a model for those patients who maintain that they cannot change their individual situations. Physiologic feedback is also important. By monitoring their levels of pain, anxiety, depression, and fatigue, persons with fibromyalgia can optimize their new levels of activity.[246,257]

In any interdisciplinary treatment model, it is necessary that the various disciplines support and appreciate each other's contributions.[250] Clinicians must move beyond the common Cartesian monistic and dualistic treatment paradigms based on mind–body separation and positivism. It is counterproductive to have the physician work from a somatogenic perspective, while the psychologist or clinical social worker considers only the psychogenic perspective. Whereas different disciplines are responsible for specific components of the overall management, treatment strategies of one discipline should be considered by other team members. Physicians and physical therapists should incorporate basic psychological techniques, medical interventions, and physical therapy procedures.[258–260] Psychologists and clinical social workers must be familiar with the goals and objectives of medicine and physical therapy. Each discipline must synchronize its efforts with any of the others.[250] Following is an overview of the role of physicians and physical therapists in the management of persons with fibromyalgia. The role of psychologists and clinical social workers is beyond the context of this chapter and will not be included. It should be obvious that the successful management of persons with fibromyalgia cannot be accomplished without mental health professionals both in group and individual interventions. Psychological group interventions may focus on problem-solving techniques, stress reduction, effective communication, and increasing the overall knowledge base, whereas individual sessions may deal with the many psychosocial issues outlined previously, including depression, anxiety, histories of sexual and physical abuse, alcoholism, illness behavior, somatization, posttraumatic stress, and so forth.[121,260–264]

Medical Management

Physicians are usually the first point of contact for the person with fibromyalgia and ultimately responsible for providing the patient with the appropriate medical diagnosis. It is common that patients with fibromyalgia have already seen multiple health care providers by the time they see a pain specialist. As discussed previously, the clinical diagnosis of fibromyalgia should be made only after excluding other causes of widespread pain. In clinical practice, it is probably irrelevant whether the patient meets the ACR research criteria. After making the diagnosis of fibromyalgia, the physician should provide patients and their families with adequate information regarding the syndrome and assist patients with developing short and long-term goals. Symptomatic and functional goals should be emphasized, rather than a cure of fibromyalgia or a total relief of pain.[26] The goals of intervention should reflect the patient's goals, needs, and desires and not necessarily the goals of the health care providers. Patients who have assisted in developing their goals are more likely to assume ownership of those goals and work toward accomplishing them with the support of health care providers. The specific goals for each discipline must support the overall goals of the patient. The interdisciplinary approach delegates responsibility back to the patients and their significant others, but the responsibility for outcomes is shared by all members of the team, including the patients.[250]

In most interdisciplinary settings, the physician is responsible for directing the rehabilitation program. The medical management includes the prescription of medications and in most cases, the referral to other disciplines, as drug therapy alone is rarely sufficient.[265] The general principles that apply to the treatment of any patient with chronic pain apply. Based on published research, there are some pharmacological interventions that appear more effective than others, although none of the medications commonly used are specific for fibromyalgia and none are very effective.[265] It is likely that new medications or combinations of medications will be used as the understanding of underlying neurochemical and hormonal pertubations increases.[54] In an era of evidence-based medicine, the pharmacological management should be based on scientific findings and subjected to clinical outcome studies.

There is some evidence that tricyclic medications may be useful. Tricyclics can improve slow-wave sleep and increase the availability of serotonin.[54,266] There are several studies that support the administration of amitriptyline (Elavil) and cyclobenzaprine (Flexeril).[267–275] Surprisingly, amitriptyline did not affect the sleep anomaly reported in some patients with fibromyalgia.[99] The recommended dose is 10–50 mg of amitriptyline and 10–30 mg of cyclobenzaprine.[265] The long-term efficacy of amitriptyline and cyclobenzaprine could not be demonstrated.[276] In another study, the combination of ibuprofen (Advil) and alprazolam (Xanax) was recommended.[277] Alprazolam is an anti-anxiety agent and is usually prescribed for the short-term relief of mild to moderate anxiety or tension. The selective serotonin reuptake inhibitors may also be of value. Goldenberg and colleagues reported improvements in sleep, pain, and overall well-being with a combination of amitriptyline and fluoxetine (Prozac). The combination of the two drugs was more effective than either drug alone.[268] Fluoxetine (20 mg) is given in the morning to avoid further insomnia.[54] Others have studied the effect of 5-hydroxytryptamine type 3 receptor antagonists and reported that both ondansetron (Zofran) and tropisetron (Navoban) significantly improved pain intensity, pain score, tender points, and average pain threshold.[278,279] Zolpidem (Ambien) was shown to have a positive effect on sleep patterns, but not on pain intensity, mood, sleep quality, morning fatigue, and the number of tender points.[280] Anti-inflammatory medications were shown to have little or no effect.[265] Biasi and colleagues reported positive results with tramadol (Ultram).[281] Bennett and colleagues tested their hypothesis that growth hormone plays a role in the generation of some of the symptoms by giving growth hormone to patients with low levels of insulin-like growth factor-1.[220,282] They observed that women with fibromyalgia and low levels of insulin-like growth factor-1 experienced an improvement in their overall symptomatology and number of tender points after nine months of daily growth hormone therapy, but no patient had a complete remission of symptoms. All patients who experienced improvement while taking growth hormone encountered a worsening of symptoms over a period of one to three months after stopping treatment.[223,282] Leal-Cerro and colleagues confirmed that the administration of growth hormone may reverse some of the symptoms of fibromyalgia.[224] The widespread use of growth hormone is, however, unrealistic because of its high cost.

Physical Therapy Management

When patients are referred to physical therapy with a medical diagnosis of fibromyalgia, the physical therapist must examine the patient and determine the appropriate physical therapy diagnosis.[283] In clinical practice, many patients diagnosed with fibromyalgia may have other treatable diagnoses as discussed previously. Typically, physical therapists are not trained to rule out medical causes of widespread pain, such as complications of cholesterol-lowering medications, hypothyroidism, or myoadenylate deaminase deficiency, but they should be able to assess patients for the presence of myofascial trigger points, hypomobility, or hypermobility. If the symptoms correlate with myofascial trigger points or with altered joint mobility, the physical therapist should review this with the referring physician and suggest that perhaps the patient may not have fibromyalgia after all. In many cases, the patient needs to be convinced that their condition may actually be treatable, which may become the main objective during the first few treatment sessions. Again, after being diagnosed with fibromyalgia, many patients modify their expectations, lifestyle, and perspectives and resort to living with a chronic incurable disease entity.

In addition to education, the most important aspect of physical therapy intervention is cardiovascular training.[265,284] Persons with fibromyalgia tend to be deconditioned.[137] Although they may perceive that exercise will worsen their condition, several studies have shown that persons with fibromyalgia can participate in regular low-intensity cardiovascular training programs without experiencing an increase in symptoms.[285–289] The physical therapist must

educate patients with fibromyalgia regarding the multiple positive effects of regular exercise on depression, quality of sleep, levels of serotonin, dehydroepiandrosterone sulphate and insulin-like growth factor-1 levels, psychological well-being, overall fitness levels, and fatigue. When comparing a program emphasizing cardiovascular training with a flexibility program, patients receiving cardiovascular training showed significantly improved cardiovascular fitness and improvements in pain threshold scores, but not in perceived pain intensity, percent body area involved, or sleep patterns.[285] Wigers and colleagues compared aerobic exercise with a stress management program and concluded that aerobic exercise was the most effective treatment approach, although there were no significant differences between the two groups at four years of follow-up.[289] Other studies also suggested that regular exercise, including aerobic walking, was correlated with less symptoms.[245,290] Nørregaard and colleagues did not find any improvement in pain, fatigue, general condition, sleep, depression, functional status, muscle strength, or aerobic capacity in either a progressive exercise program or an aerobic dance program, partly due to poor compliance.[291] A common problem with any form of exercise is the lack of consistent long-term compliance. Whenever untrained individuals start to exercise, they will experience an initial increase of muscular pain, not to be confused with the typical pain associated with fibromyalgia. The skill in treating patients is the appropriate timing and coordinating of various aspects of rehabilitation. Each patient has a distinct personality, lifestyle, and activity level that need to be considered during the rehabilitation process. Will the patient be successful in undertaking a home program? Will the patient be overly zealous in the early aspects of strength or cardiovascular training? A gradual adaptation to a progressive exercise program is usually well tolerated and may include lower or upper body ergometry, walking, or aquatic physical therapy.[54] Along with cardiovascular training, light strength training is appropriate. Strength training should be approached with some caution. Free weights should be avoided, unless very small hand weights are being used; controlled weight machines are preferred. Strength training is meant not to replace cardiovascular training, but to complement it. The patient is encouraged to stretch before and after workout to maintain flexibility. Other approaches may be considered as well, including the Feldenkrais Method, the Alexander Technique, T'ai Chi, or yoga, although there are no scientific studies specific for fibromyalgia and these somatic approaches.[292]

Soft tissue restrictions and joint hypomobility should be assessed and corrected when indicated, realizing that these restrictions are most likely the result of decreased activity levels and not involved in the etiology of fibromyalgia. Müller and colleagues speculated that spinal hypomobility may be involved in the etiology of fibromyalgia, but did not offer convincing support for this notion.[293] There are no studies that support the use of myofascial or joint manipulations, although a correlation was established between perceived pain severity and physical functioning, defined by spinal mobility tests.[294] Acupuncture and electro-acupuncture were effective in relieving the symptoms of fibromyalgia, although the long-term results have not been studied yet.[295,296] Gunn suggested using the dry needle technique of intramuscular stimulation, but has not completed any prospective studies on the effects of intramuscular stimulation on the symptoms of fibromyalgia.[297]

Taxonomy

As the most widely accepted model for the pathogenesis of fibromyalgia suggests hypersensitivity of the central nervous system and a dysfunctional endocrine system, rather than pathologically painful muscles, the question emerges whether fibromyalgia should still be considered a "muscle pain syndrome."[298] Even the name "fibromyalgia" no longer seems appropriate for the syndrome, as it suggests that muscles and fibrous tissues are causally involved in the generation of pain. Furthermore, it may suggest that pain is limited to fibrous tissues and muscles.

Because persons with fibromyalgia display a generalized, decreased pain threshold, Russell suggested that fibromyalgia can be considered "chronic widespread allodynia," as it meets the criteria for allodynia as defined by the International Association for the Study of Pain.[299,300] Allodynia is defined as "a painful response to a normally non-painful stimulus."[299] This modified descriptor of fibromyalgia does not consider the multiple other features of the diagnosis, including hypervigilance, hyperalgesia, psychosocial dysfunction, and so forth. Considering all the different aspects of the syndrome, perhaps a more appropriate name is "complex widespread-pain syndrome," analogous to the development of the term "complex regional painsyndrome," which replaced the misleading terms "reflex sympathetic dystrophy" and "causalgia."[298,301]

MYOFASCIAL PAIN SYNDROME

Definition

Myofascial pain syndrome has been defined differently by different authors or disciplines. Sometimes, myofascial pain syndrome is defined as a regional pain syndrome of any soft tissue origin.[302] In dentistry, myofascial pain dysfunction syndrome has become the commonly used term, described as muscle pain with or without limitations in mouth opening.[303] Gunn described myofascial pain syndrome as "chronic pain conditions that occur in the musculoskeletal system when there is no obvious injury or inflammation."[304] The most commonly used definition of myofascial pain syndrome is formulated by Simons, Travell, and Simons as a muscle pain disorder characterized by the presence of a myofascial trigger point within a taut band, local tenderness, referral of pain to a distant site, restricted range of motion, and autonomic phenomena.[305] Autonomic phenomena may include vasoconstriction, pilomotor response, ptosis, and hypersecretion.[306] Simons, Travell, and Simons have described myofascial trigger points in almost all skeletal muscles of the body.[305,307] Trigger points can be present in muscle, skin, fascia, ligaments, joint capsule, and periosteum; however, nearly all research has focused on muscle trigger points.[305] In the European literature, the term "myogelosis" is commonly used instead of "myofascial trigger point."[7]

Although in clinical practice, the Simons, Travell, and Simons criteria appear to be acceptable, the criteria have not been subjected to scientific research and lack established reliability and validity. During the 1998 Fourth World Congress on Myofascial Pain and Fibromyalgia in Italy, the International Myopain Society established a multidisciplinary committee to design a study model for validation of the diagnostic criteria. The committee aims to establish reliable methods for diagnosis of myofascial pain syndrome, determine the interrater reliability of trigger point examination, and determine the sensitivity and specificity with which classification criteria can distinguish patients with myofascial pain syndrome from healthy control subjects.[308] In this chapter, the Simons, Travell, and Simons criteria are applied.

Myofascial pain syndrome can be acute in nature or become a persistent chronic pain problem.[309] It has been reported as the most common diagnosis responsible for chronic pain and disability.[310–312] Myofascial trigger points are found equally in men and women and are commonly found in children.[305,313] Myofascial pain syndrome is often thought of as a regional pain syndrome in contrast to fibromyalgia as a widespread syndrome. Data by Gerwin revealed that as many as 45% of patients with chronic myofascial pain have generalized pain in three or four quadrants.[314,315] Although these patients may also meet the ACR criteria for fibromyalgia, they featured myofascial trigger points within taut bands as the main source of their pain, making myofascial pain syndrome the preferred diagnosis. Myofascial pain syndrome can exist in isolation without involvement of other structures, or be associated with other musculoskeletal disorders, including facet joint injuries, disc herniations, osteoarthritis, or as part of post-laminectomy syndromes. It can occur as a complication of certain medical conditions,

including myocardial infarction or kidney disorders. Myofascial pain syndrome should be considered in the differential diagnosis of radiculopathies, anginal pain, joint dysfunction (including craniomandibular dysfunction), migraines, tension headaches, complex regional pain syndrome, carpal tunnel syndrome, repetitive strain injuries, whiplash injuries, and most other pain syndromes.[310,316–325] Myofascial pain resulting from muscular dysfunction is called primary myofascial pain. In secondary myofascial pain syndrome, the pain and muscle dysfunction are the result of underlying medical pathology, joint or mechanical dysfunction, or psychological dysfunction. Hendler and Kozikowski concluded that primary and secondary myofascial pain were the most commonly missed diagnoses in chronic pain patients. A thorough diagnostic evaluation was recommended to identify the underlying myofascial cause of chronic pain, rather than considering the pain problem to be psychogenic in nature.[326,327] From a practical perspective, there is no diagnostic or clinical benefit to the patient in making the distinction between primary and secondary myofascial pain syndrome.

The concept of primary and secondary myofascial pain syndrome was questioned by Quintner and Cohen, who instead deemed all myofascial pain syndrome phenomena the result of secondary hyperalgesia of peripheral neural origin.[328] Gunn hypothesized that symptoms of myofascial pain are always secondary to neuropathies, especially radiculopathies. By applying Cannon and Rosenblueth's law of denervation, Gunn concluded that myofascial pain is the result of functional or structural alterations within the central and peripheral nervous system.[304] According to Cannon and Rosenblueth's law of denervation, nerves and their innervated structures develop "supersensitivity" when the nerves are not functioning properly.[329] Gunn described that the autonomic phenomena, including vasomotor, sudomotor, and pilomotor changes, are features of the neuropathy model and not specifically of myofascial trigger points.[304]

Diagnosis

The main criterion for the diagnosis of myofascial pain syndrome is the presence of an active myofascial trigger point, an exquisitely sensitive region in a taut band of skeletal muscle consisting of multiple sensitive trigger loci.[330,331] Most patients complain of more global, diffuse pain and are not aware that specific myofascial trigger points may cause their pain. The key features of the trigger point have been established by Simons, Travell, and Simons and are listed in Table 6–2.[305]

The diagnosis of myofascial pain syndrome is made by systematic palpation of taut bands and myofascial trigger points, following a review of the patient's history, and an assessment of

---

Table 6–2 Criteria for Identifying a Myofascial Trigger Point

Essential criteria

1. Taut band palpable (if muscle is accessible).
2. Exquisite spot tenderness of a nodule in a taut band.
3. Patient's recognition of current pain complaint by pressure on the tender nodule (identifies an active trigger point).
4. Painful limit to full stretch range of motion.

Confirmatory observations

1. Visual or tactile identification of local twitch response.
2. Imaging of a local twitch response induced by needle penetration of tender nodule.
3. Pain or altered sensation (in the distribution expected from a trigger point in that muscle) on compression of tender nodule.
4. Electromyographic demonstration of spontaneous electrical activity characteristic of active loci in the tender nodule of a taut band.

Source: Reprinted with permission from D.G. Simons, J.G. Travell, and L.S. Simons, *Myofascial Pain and Dysfunction: The Trigger Point Manual 2/E*, Vol. 1, Lippincott Williams & Wilkins, © 1999.

---

posture and functional movement patterns.[305] The patient's pain pattern and range-of-motion restrictions usually point the clinician to the involved muscles. According to Gerwin and colleagues, the minimum criteria that must be satisfied in order to distinguish a myofascial trigger point from any other tender area in muscle are a taut band and a tender point in that taut band. The presence of a local twitch response, referred pain, or reproduction of the person's symptomatic pain increased the certainty and specificity of the diagnosis of myofascial pain syndrome.[332] Systematic palpation will differentiate between myofascial taut bands and general muscle spasms.[333] Spasms can be defined as electromyographic activity as the result of increased neuromuscular tone of the entire muscle. A taut band is a localized contracture within the muscle without activation of the motor endplate.[334] The taut band, trigger point, and local twitch response are objective criteria, identified solely by palpation, that do not require a verbal response from the patient. A local twitch response is an indication of the presence of an active trigger point. It is a brief, involuntary contraction of the taut band that can be recorded electromyographically, be felt with the needle during trigger point injection or needling, or observed visually or on diagnostic ultrasound. It is mediated primarily through the spinal cord without supraspinal influence.[330,335] The patient's body type and specific muscle determine the ease of soliciting a local twitch response.

The interrater reliability of the myofascial trigger point examination has been studied by several authors; however, it was only recently established by Gerwin and colleagues for the five major features of the trigger point.[332,336–339] Even in this study, a team of recognized experts could initially not agree. Only after developing consensus regarding the criteria, did the experts agree, which indicates that training is essential for the identification of myofascial trigger points. Gerwin and colleagues established that individual features of the trigger point are differentially represented in different muscles. For example, the local twitch response was easier to obtain and, therefore, more commonly found in the extensor digitorum communis than in the infraspinatus muscle.[332]

The degree of stimulation required to reproduce a patient's usual pain determines whether a trigger point is considered active or latent.[340] An active trigger point has a lower pain threshold than a latent trigger point. A trigger point is considered active when normal physiological movements or postures cause pain, whereas a latent trigger point requires a significant amount of mechanical stimulation to reproduce pain. Various authors have suggested methods to objectively quantify the amount of pressure required to elicit a painful response from a trigger point using algometry or palpometry; however, it remains difficult to determine the distinguishing features of active and latent myofascial trigger points.[341,342] It is important to realize that pressure algometry is influenced by nociceptors in the skin and subcutaneous tissues.[343]

Both active and latent myofascial trigger points may cause dysfunction, including restrictions in range of motion and muscle weakness.[321] In patients with acute myofascial pain, restrictions in range of motion are primarily due to shortening of muscle fibers, pain, and kinesiophobia. In chronic cases, soft tissue and joint adhesions can further contribute to restrictions in range of motion.[344] Muscle weakness without atrophy is often seen with myofascial pain syndrome. Muscle weakness may be due to pain, restrictions in range of motion, kinesiophobia, inhibition of gamma motoneuron activity, or reflex inhibition of anterior horn cell function as a result of painful sensory input.[345,346] Activation of the trigger point can produce several autonomic phenomena (i.e., vascular effects, changes in skin temperature, and secretory, pilomotor, and trophic changes). Trophic changes may lead to the development of so-called "satellite trigger points" in the area of referred pain.[305] Gunn considered the trophic changes essential to the diagnosis and treatment of neuropathy.[304] Autonomic changes are not specific for myofascial pain syndrome, as most pain syndromes have an autonomic component.[347]

The diagnostic process must include the usual differential diagnostic considerations, and rule out other pathological processes. For example, in the examination of a patient with knee pain, the clinician should consider ligamentous, meniscal, and capsular injuries, patellofemoral joint dysfunction, bursitis, tendinitis, and arthritis, but also appreciate referred pain patterns and the biomechanical implications of taut muscle bands and myofascial trigger points in the quadriceps, hamstrings, gluteals and iliotibial band, adductors, and calf muscles.[348] After establishing the initial diagnosis of myofascial pain syndrome, the clinician must determine any mechanical, systemic, or psychological perpetuating factors that may contribute to the formation or persistence of myofascial trigger points. Major mechanical factors to be considered in the diagnosis and management of myofascial pain syndrome include anatomic variations and poor postures. Myofascial trigger points and taut bands may also contribute to further mechanical dysfunction.

Mechanical dysfunction is one of the main problems of myofascial pain. Correcting mechanical dysfunction has become the main objective of Gunn's intramuscular stimulation approach to myofascial pain syndrome.[304] Physical therapists may use soft tissue mobilization as well to correct mechanical dysfunction. For example, considering that knee joint motion is accompanied by simultaneous coactivation of the quadriceps and hamstrings muscles, any mechanical discrepancy in either muscle group will affect the resultant joint motion and possibly influence joint stability. It is conceivable that a taut band in the semimembranosis muscle restricts the mobility of the medial and, perhaps, even the lateral meniscus through its insertions. The semimembranosis muscle reinforces the posteromedial aspect of the knee capsule. It can flex and internally rotate the tibia on the femur and pull the posterior horn of the medial meniscus posteriorly during flexion of the knee.[349] Perhaps, a semimembranosis muscle shortened by taut bands and myofascial trigger points maintains the menisci in a relative posterior position even during extension of the knee. Myofascial trigger points in the semimembranosis muscle may, therefore, increase the likelihood of meniscal injury.[348] In addition to treating the local muscles, Gunn advocated evaluating and treating the paraspinal muscles at the levels of segmental innervation, including L2–3, L3–4, and L4–5.[304] Following in Gunn's footsteps, Fischer also promoted treatment of the paraspinal muscles, as well as the supraspinous and interspinous ligaments.[350] Where Gunn recommended dry needling of the multifidi muscles, Fischer recommended lidocaine injections into the spinal ligaments.[304,350]

Systemic medical factors that can interfere with recovery from myofascial pain syndrome are medical conditions that either affect the muscle energy system or otherwise interfere with muscle metabolism.[305] Commonly seen conditions include iron, folic acid, and vitamin B_{12} insufficiencies and hypothyroidism.[314,320,351] Less common systemic factors are gout, hypercalcemia, and infections, including recurrent yeast infections and amoebiasis; however, there are no epidemiologic studies supporting these clinical observations.[320] Psychological perpetuating factors may include depression, anxiety, stressful life circumstances, anger, and hopelessness. Patients with myofascial pain syndrome were reported to have significantly more severe depression and more problems with social relationships than patients with certain other pain syndromes, such as arthritis.[352–354]

Some authors have questioned the validity of myofascial pain syndrome or its underlying mechanisms.[127,328,355–357] During the past few years, several objective features have been described in the scientific literature that further substantiate the existence of myofascial trigger points. Several researchers established that trigger points have a specific electrical discharge characteristic when using needle electromyography. Indwelling electromyography, however, does not replace manual palpation and does not add any significant value to the clinical diagnostic process.[358–361] Surface electromyography can be valuable for identifying muscle asymmetries

and dysfunctional muscle firing patterns. It does not demonstrate the electrical activity of myofascial trigger points, however, or confirm the diagnosis of myofascial pain syndrome.[362–365] There are no laboratory or imaging studies available for the diagnosis of myofascial pain syndrome or myofascial trigger points. High resolution sonography was not sensitive enough to visualize the actual trigger point, but allowed researchers to visualize the twitch response of the taut band following stimulation of the trigger point by insertion of a hypodermic needle.[366,367]

Clinical Characteristics

Clinically, patients usually complain of diffuse pain confined to one or more regions of the body, as opposed to fibromyalgia, which always features widespread pain. In some instances, patients describe sharp pain over myofascial trigger points that they can easily identify. In other cases, the pain complaint is related to referred pain from myofascial trigger points. On palpation, taut bands and myofascial trigger points can be identified. As many chronic pain patients perceive that clinicians do not always take them and their pain complaints seriously, eliciting the familiar pain complaint by compressing myofascial trigger points may surprise patients initially, but may be an important step in establishing a therapeutic relationship. Myofascial pain syndrome may follow a sequence of events associated with postural imbalances, leading to joint hypermobility or hypomobility and abnormal biomechanical functioning. A prime example is the forward head posture, a common precursor to myofascial pain syndrome. Patients with chronic myofascial pain syndrome may report other symptoms associated with pain, such as sleep disturbances, fatigue, and increased irritability.

Referred Pain

An active myofascial trigger point refers pain usually to a distant site. The referred pain pattern is not necessarily restricted to single segmental pathways or to peripheral nerve distributions.

Although typical referred pain patterns have been established, there is considerable variation in between patients.[305,307] Usually, the pain in reference zones is described as "deep tissue pain" of a dull and aching nature. Occasionally, patients may report burning or tingling sensations.[305,368–370] By mechanically stimulating an active trigger point, patients may report the reproduction of their pain, either immediately or after a 10- to 15-second delay. Mechanical stimulation can consist of manual pressure, needling of the trigger point, movement of the involved body region, and postural strains, such as forward head posture or pressure on the gluteal muscles in sitting. Even physiological muscle tone at rest may stimulate an active trigger point, which is indicative of hypersensitivity of the nervous system. Normally, skeletal muscle nociceptors require high intensities of stimulation and they do not respond to moderate local pressure, contractions, or muscle stretches.[334,371] Referred pain is not specific to myofascial pain syndrome; however, it is more common and much easier to elicit over myofascial trigger points.[340] Normal muscle tissue and other body tissues may also refer pain to distant regions with mechanical pressure (i.e., the skin, zygopophyseal joints, or internal organs), making referred pain elicited by stimulation of a tender location a nonspecific finding.[306,370,372–376] Gunn no longer considers referred pain an essential feature of myofascial pain syndrome, which has become one of the differences between Gunn's diagnostic and treatment approach and Simons, Travell, and Simons' approach.[305,377] Referred pain is no longer considered a diagnostic symptom but can guide a clinician to determine which muscles have active myofascial trigger points (Figures 6–4 and 6–5).

Anatomic Variations

Many persons with myofascial pain syndrome feature anatomic variations that may contribute to myofascial trigger point formation. It is not unusual that a particular anatomic variation did not cause any dysfunction prior to the event that resulted in the onset of myofascial pain syndrome, yet became a significant factor during the recovery. For example, a patient with a significant leg length discrepancy may never have had low back pain; however, following a motor vehicle accident, the discrepancy may become a critical perpetuating factor for myofascial trigger points in the quadratus lumborum muscle. Gunn maintained that this is due to an already supersensitive peripheral nervous system. The added stress of a motor vehicle accident may exceed the patient's threshold and result in complaints of persistent pain.[304] According to Simons, Travell, and Simons, the most common anatomic variations are leg length discrepancy, small hemipelvis, short upper arm syndrome, and long second metatarsal syndrome.[305,307] Leg length discrepancies may be due to congenital, developmental, traumatic, or pathological changes in one of the osseous links of the lower extremity kinetic chain. A distinction must be made between a structural and a functional leg length discrepancy. Structural discrepancies are due to true anatomic differences in length of the femur or tibia, whereas functional discrepancies can be caused by hip adductor contractures, hip capsule tightness, or by unilateral innominate rotation. Leg length discrepancies and pelvic asymmetries may produce muscle imbalances and postural adjustments and result in the development of myofascial trigger points.[378] Short upper arms result in pronated shoulders, pectoral muscle shortening, and abnormal loading of neck and trunk muscles, as the individual attempts to find a comfortable position when seated. Another cause of biomechanical stress on muscle that can lead to persistent myofascial trigger points is a long second metatarsal bone. In this situation, the normal, stable tripod support of the foot created by the first and second metatarsal bones anteriorly, and the heel posteriorly, may not occur. Instead, in some individuals with this foot configuration, weight is carried on a knife-edge from the second metatarsal head to the heel, overloading the peroneus longus. Diagnostic callus formation occurs in these individuals in the areas of abnormal loading, under the second metatarsal head, and on the medial

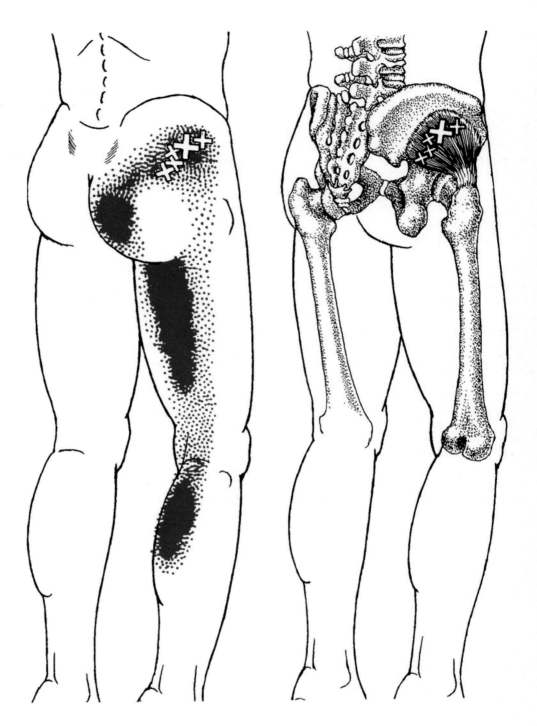

Figure 6–4 Referred pain patterns of the gluteus minimus muscle mimic sciatic nerve pain. *Source:* Reprinted with permission from *Mediclip, Manual Medicine 2*, version 1.0a., Williams & Wilkins.

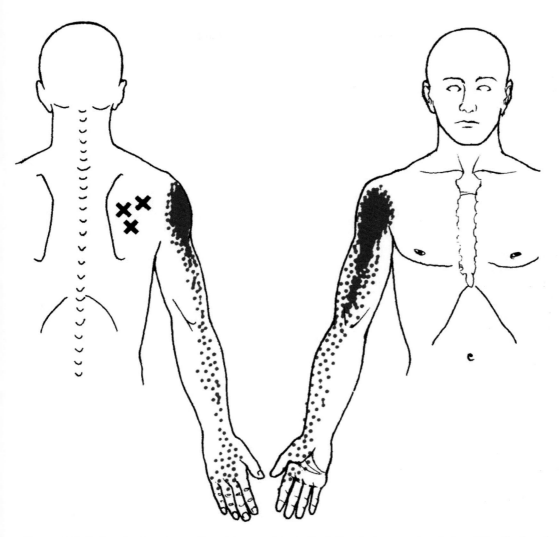

Figure 6–5 Referred pain patterns from trigger points in the infraspinatus muscle mimic a C6 radiculopathy. *Source:* Reprinted with permission from *Mediclip, Manual Medicine 2*, version 1.0a., Williams & Wilkins.

aspect of the foot at the great toe and first metatarsal head.[320] Although there is still considerable controversy regarding the biomechanical implications of poor occlusion on the development of myofascial trigger points in the craniomandibular muscles, it is likely that occlusal problems, including missing teeth and early contacts, contribute to mechanical stress on muscles

and their associated pain problems of headaches, tooth, and facial pain.[379,380]

Posture

Abnormal postures can result in muscle imbalances, the formation of myofascial trigger points in adaptively shortened or lengthened muscles, joint hypomobility and hypermobility, and nerve compression. Forward head posture is the most common postural deviation in chronic pain patients, including patients with myofascial pain syndrome.[378,381] The biomechanical and myofascial aspects of the forward head posture are fully discussed in Chapter 7. The typical symptoms in this particular scenario (Table 6–3) can include:

- Intermittent cervical, thoracic, or lumbar pain
- Unilateral or bilateral headaches and facial pain
- Myofascial trigger points in multiple muscle sites
- Upper extremity referred pain or paresthesia in the absence of neurological findings
- Difficulty sitting for a long period of time, especially in deep, soft chairs or bucket seats that accentuate forward-head posture

Table 6–3 Postural Problems Found in 164 Patients with Myofascial Pain Syndrome of the Head and Neck

	N	%
Body		
Poor sitting/standing posture	157	96.0
Forward head tilt	139	84.7
Rounded shoulders	135	82.3
Poor tongue position	111	67.7
Abnormal lordosis	76	46.3
Scoliosis	26	15.9
Occlusion		
Slide from retruded contact position to intercuspal contact position of 1 mm or greater	140	85.5
Unilateral occlusal prematurities in intercuspal contact position	113	68.9
Class II, Division 1	96	58.5
Class II, Division 2	51	31.1
Class III	16	9.8

Source: Reprinted with permission from J.R. Fricton, Myofascial Pain Syndrome: Characteristics and Clinical Epidemiology, *Advances in Pain Research and Therapy,* Vol. 19, p. 121, © 1989, Lippincott Williams & Wilkins.

- Pain or ache on prolonged standing
- Pain decreased by rest or gentle movements

Several studies have shown that occupational groups with constrained work postures and repetitive arm movements are at increased risk for developing myofascial pain syndrome.[382,383] Work tasks with high repetition frequency and static muscle loading may actually decrease the pain pressure threshold and result in allodynia and hyperalgesia.[384] Awkward postures are common in the workplace and include excessive wrist flexion and extension, ulnar and radial abduction, forearm supination and pronation, extended reaches beyond the shoulder-reach envelope, and pinch grips that are either too wide or too narrow.[385,386] Skubick and colleagues demonstrated that asymmetrical loading of the sternocleidomastoid muscles and cervical paraspinal muscles can result in carpal tunnel syndrome.[387] Particular occupational groups at increased risk include musicians, data entry operators and typists, industrial workers, and assembly line workers.[388–392] Andersen and colleagues reported the onset of myofascial pain syndrome in various occupational groups with monotonous repetitive work.[382] In a study of patients with cumulative trauma disorders, 94.5% were diagnosed with myofascial pain syndrome.[383]

Pathogenesis

Musculoskeletal Abnormalities

There is some evidence of histologic changes at the site of myofascial trigger points identifiable by light microscopy.[334] In 1951, Glogowsky and Wallraff reported damaged fibril structures in "myogeloses." In later studies, Fassbender observed degenerative changes of the I-bands, in addition to capillary damage, a focal accumulation of glycogen, and a disintegration of the myofibrillar network.[393–395] In 1995, Gariphianova described pathological changes with biopsy studies of myofascial trigger points, including a decrease in quantity of mitochondria, possibly indicating metabolic distress.[396] Reitinger and colleagues also reported pathologic alterations of the mitochondria, as well as increased A-bands and decreased I-bands in muscle sarcomeres of myofascial trigger points in the gluteus medius muscle; however, they did not describe their definition of a trigger point.[397] Pongratz and Späth noticed segmental degeneration of muscle fibers with concomitant edema and histiocytic cellular reaction.[398]

Energy Crisis Hypothesis

Both the local tenderness and taut bands characteristic of myofascial pain syndrome are proposed to be associated with the "energy crisis hypothesis."[305] This hypothesis postulates that there is decreased circulation and local ischemia in a myofascial trigger point due to sustained sarcomere shortening. Studies by Brückle and colleagues, measuring extremely low oxygen levels (5% of normal) within myofascial trigger points, appeared to confirm the hypoxia component of the energy crisis hypothesis.[399] The shortening of the actin-myosin complex can be caused by a traumatic release of calcium either from the sarcoplasmic reticulum or from a failure to restore adenosine triphosphate. The possible roles of titin and nebulin have not yet been considered in the etiology of myofascial trigger points. Adenosine triphosphate is essential for normal functioning of the calcium pump, as well as for the release of the actin-myosin complex. A shortage of adenosine triphosphate can result in local muscle contractures or taut bands.[334] The pathologic alterations of the mitochondria can further contribute to a shortage of adenosine triphosphate. Termination of a muscle contraction is normally accomplished by pumping calcium back into the sarcoplasmic reticulum against a large concentration gradient. With an impaired calcium pump, the intracellular calcium concentration stays elevated, and the actin and myosin filaments become continuously activated.[400] Shenoi and Nagler confirmed that an impaired reuptake of calcium into the sarcoplasmic reticulum can cause myofascial trigger points. They reported that calcium channel

blockers caused myofascial trigger points, presumably based on their ability to prevent calcium re-uptake.[401]

Electrophysiologic Abnormalities

In 1957, Weeks and Travell published a report that outlined a characteristic electrical activity of a myofascial trigger point.[358] It was not until 1993 that Hubbard and Berkoff confirmed the presence of specific electromyographic activity in myofascial trigger points of the trapezius muscle. They observed that this activity was greater than the electromyographic activity in a nontender area of the same muscle. They recorded both low amplitude continuous action potentials and intermittent spikes from active myofascial trigger points.[359] Simons and colleagues reported similar action potentials of 10 to 50 μV, which they defined as "spontaneous electrical activity," in contrast to the intermittent biphasic spikes of 100 to 600 μV.[360,361] The electrical activity is not mediated through the spine or supraspinal influences, suggesting that it may be a motor endplate phenomenon. The electrical activity was found to be similar to abnormal endplate potentials, associated with an excessive release of acetylcholine, which affects the voltage gated sodium channels of the sarcoplasmic reticulum and increases the intracellular calcium levels.[306,402–404] Gunn articulated that the relative increase of acetylcholine release into the muscle may be the result of neural dysfunction, associated with a decrease of the available acetylcholinesterase and the renewed activation of acetylcholine receptors throughout the muscle.[304] It is not clear whether there are, in fact, newly formed acetylcholine receptors.[405,406]

Several studies have demonstrated that myofascial trigger points are nearly always located in the region of the motor endplate zone.[361,407] Hong proposed that a palpable myofascial trigger point consists of multiple discrete sensible loci. Each locus may contain one or more sensitized nociceptive nerve endings. It is likely that these spots represent abnormal motor endplates.[407,408] In summary, myofascial trigger points are probably associated with dysfunctional motor endplates.[305] The finding that injections with botulinum toxin are effective in inactivating myofascial trigger points further supports the motor endplate hypothesis.[409-414] Botulinum toxin is a neurotoxin that blocks the release of acetylcholine from presynaptic cholinergic nerve endings. A recent study in mice demonstrated that the administration of botulinum toxin resulted in a complete functional repair of the dysfunctional endplates.[415]

Autonomic Contributions

Based on the finding that the electromyographic activity of myofascial trigger points increased as the result of psychological stress, Hubbard and colleagues proposed that myofascial trigger points are associated with the autonomic nervous system.[416-418] Autonomic phenomena have always been described as part of myofascial pain syndrome.[305] Several studies have now shown that the administration of the sympathetic blocking agent phentolamine significantly reduces the electrical activity of a myofascial trigger point, which supports the hypothesis that the autonomic nervous system is involved in the pathogenesis of myofascial trigger points.[418,419] In an uncontrolled biopsy study, Hubbard identified a single muscle spindle at the site where the spontaneous electrical activity was recorded.[418] As the muscle spindle is autonomically innervated, Hubbard proposed that myofascial trigger points are associated with dysfunctional muscle spindles. Partanen supported this notion by expressing that, in his opinion, the endplate spikes are indeed action potentials of intrafusal muscle fibers and that the "active spots" are in fact muscle spindles. Simons and colleagues refuted this, however, by demonstrating that the spike potentials are propagated by extrafusal muscle fibers and not by intrafusal fibers.[420,421]

At this point, the available data are inconclusive. The motor endplate studies clearly support the hypothesis that myofascial trigger points are dysfunctional motor endplates, whereas other studies support the role of the sympathetic nervous system. The two concepts are not mutually exclusive. There is, however, little evidence that the effect of the autonomic nervous system on myofascial trigger points is applied via the muscle spindle. It is conceivable that, due to the constant increased stress within a taut band, the muscle spindle is exposed to static loading, which may result in a steady discharge of impulses, known as the static response of the muscle spindle. This would not explain why the administration of phentolamine would reduce the electrical activity of the myofascial trigger point. Static stress applied to the muscle spindle may lengthen the equatorial part of the intrafusal muscle fibers; however, that would still not explain the formation of myofascial trigger points. The mechanism of the interactions between the autonomic nervous system and myofascial trigger points needs further investigations. Direct connections between the sympathetic nervous system and muscle fibers have been established and may be critical for future studies.[422] Already in 1981, Barker and Saito demonstrated that an autonomic innervation is present to some extrafusal muscle fibers.[423] Recently, Ljung demonstrated that the extensor carpi brevis muscle is supplied with heterogeneously distributed sympathetic and sensory innervations in relation to small blood vessels.[424] Assuming that other striated muscles have similar sympathetic nerve distributions, perhaps these sympathetic fibers can influence the contractibility of muscle fibers or alter the function of the motor endplate, especially under pathological conditions. It is unlikely that the sympathetic influence on muscle receptors has any functional significance under physiological conditions, but under pathological conditions, these sympathetic nerve endings may become sensitized by neuro-active substances released in the vicinity of the endplates.[305,425] More research is needed.

Central Sensitization

As with fibromyalgia, knowledge from the pain sciences must be considered. Local tenderness of myofascial trigger points is due to peripheral sensitization of nociceptors as well as neuroplastic changes within the spinal dorsal horn.[368] Vecchiet and colleagues have described specific sensory changes over myofascial trigger points. They observed significant lowering of the pain threshold over active trigger points when measured by electrical stimulation, not only in the muscular tissue, but also in the overlying cutaneous and subcutaneous tissues. This is in contrast with their findings on fibromyalgia tender points. In fibromyalgia, hyperalgesia in all three tissues was present not only over fibromyalgia tender points, but also in other nonpainful regions. With latent myofascial trigger points, the sensory changes did not involve the cutaneous and subcutaneous tissues.[66,426,427] Afferent barrage from muscles, joints, skin, and viscera can result in central sensitization by the unmasking of "sleeping" receptors.[334,428,429] Bendtsen and colleagues also suggested nociceptive reception by low-threshold mechanosensitive neurons.[430] The afferent input from these newly effective receptors may result in spatial summation in the dorsal horn and the appearance of new receptive fields. This means that input from previously ineffective regions can now stimulate the neurons.[334,431] Allodynia of a myofascial trigger point is explained by these changes in the dorsal horn, whereas hyperalgesia is the result of both peripheral sensitization and dorsal horn changes.[430,432,433] It is interesting that Gunn maintained that myofascial pain syndrome is not dependent on nociceptive input. According to Gunn, the symptoms of myofascial pain syndrome are explained entirely by the functional deficiencies of the peripheral nervous system. It appears that Gunn's hypothesis falls short in this respect, as several studies have identified the nociceptive nature of active myofascial trigger points.[66,368,426,427,429,430]

The unmasking processes of interneurons of the dorsal horn are the pathophysiological basis of the modified convergence projection theory proposed by Mense.[428] After identifying an original receptive field of the biceps femoris muscle of a rat, Mense injected a painful dose of bradykinin in the tibialis anterior muscle. Bradykinin levels have been shown to increase during pain, ischemia, static muscle contractions, and inflammation. According to Mense, the effects of bradykinin injections on the dorsal horn have similarities with the effects of myofascial trigger points.[428] The activity of the neuron corresponding with the receptive field was measured by an electrode placed in the spinal cord. After 5 minutes, the original receptive field had expanded; after 15 minutes, the original receptive field no longer responded to just painful stimuli, but also to moderate mechanical stimuli. As the interneurons are located over various segments, pain may be experienced in regions outside the segmental innervation of the myofascial trigger point, which distinguishes Mense's hypothesis from the conventional convergence theory.[434] This mechanism may result in the formation of satellite trigger points in the area of the enlarged receptive field. The temporal delay of the onset of referred pain would be the result of the time needed to unmask the interneurons with substance P and glutamate. It is likely that a similar process exists for craniomandibular muscles, even though they do not receive input from the spinal cord, as new or enlarged receptive fields were also identified after injection of mustard oil in the masseter muscle.[435] Although Mense emphasized that data from animal research may not be fully applicable to the clinical patient with myofascial trigger points, the modified convergence projection theory offers a conceivable model for the referred pain phenomena seen in myofascial pain syndrome.[428]

Another theory to explain referred pain is based on branching of primary afferent neurons in the periphery innervating both muscle and viscera. Input from one branch could activate the other branch antidromically. McMahon and Wall offered evidence of peripheral branching by recording different conduction velocities in single fibers excited distally.[436,437] In addition, the neuronal branches have been identified histologically. The theory would fall short in explaining the sensation of deep muscle pain because one of the neuronal branches terminates in the skin. The theory would also not explain referred pain in a distant location, as the neuronal branches would physically not be long enough to reach such locations.[428]

Management of Myofascial Pain Syndrome

The goals of treatment of myofascial pain syndrome are restoration of normal tissue mobility by inactivating myofascial trigger points and return to function.[320] As with any treatment plan addressing pain problems, the treatment plan can be divided into a pain-control phase and a training or conditioning phase.[438] During the pain-control phase, inactivation of the myofascial trigger points is the main short-term goal. It is important to improve the circulation at the site of the myofascial trigger point, to decrease pathological nociceptive activity, and to eliminate the abnormal biomechanical force patterns generated by the taut bands. Invasive techniques are dry needling, including intramuscular stimulation, and direct injection of a local anesthetic, saline, or botulinum A toxin into the muscle.[320] Steroid injections are not recommended for myofascial trigger points as they may induce myopathy.[365] Invasive techniques are not without risks and require thorough knowledge of anatomy, indications, and contraindications.[439,440] Injections can only be performed by physicians. Dry needling falls well within the scope of practice of physical therapy. Dry needling is a form of "mechanical stimulation," which in most physical therapy state laws is described as one of the mechanisms of physical therapy practice. Some physical therapy state laws do not allow dry needling as physical therapists are not allowed to penetrate the skin. Recently, the Maryland Board of Physical Therapy Examiners ruled that dry needling is indeed part of physical therapy practice. In clinical practice, the combination of Gunn's intramuscular stimulation with Simons, Travell, and Simons' trigger point therapy appears to be especially effective, although clinical studies have not been completed. Gunn emphasized correcting the biomechanical aspects of taut bands by needling the taut bands in muscles combined with needling of tight paraspinal muscles at the same segmental levels. Gunn no longer considers the referred pain patterns of myofascial trigger points.[304] Needling the taut band only does not necessarily eliminate active myofascial trigger points. Therefore, in addition to needling the taut band near the myotendonal junction, clinicians must inactivate myofascial trigger points either by needling or by noninvasive means. The myofascial trigger point is primarily responsible for the referred pain patterns seen in myofascial pain syndrome. There is evidence that needling trigger points in one muscle group may eliminate trigger points in muscles that belong to the referred pain area of the treated trigger points.[441] Noninvasive techniques include manual therapy, relaxation training, therapeutic exercise, the use of electrotherapy modalities, correction of structural or mechanical stressors, and resolution of possible underlying medical disorders that predispose to the development or maintenance of myofascial trigger points.[442] Banks and colleagues demonstrated recently that autogenic relaxation training reduced the electrical activity of myofascial trigger points significantly.[443]

Manual therapy is one of the basic treatment options for myofascial pain syndrome. The practitioner must evaluate and, when indicated, treat both soft tissue and joint dysfunctions. Because myofascial trigger points may be directly related to underlying articular dysfunction both in primary and secondary myofascial pain syndrome, dysfunction of muscles and joints should be considered as a single functional unit.[378,444] Soft tissue mobilization is probably the most important manual therapy component of the treatment program. In addition to treating the actual myofascial trigger points, the intratissue and intertissue mobility of the functional unit must be evaluated and treated as well. Effective soft tissue techniques include myofascial manipulation, massage therapy techniques, sustained pressure over the myofascial trigger point, stretch and spray techniques combined with post-isometric relaxation, or muscle energy/hold-relax techniques.[305,307]

Correcting structural and functional discrepancies may include specific muscle stretches, neurodynamic mobilizations, joint mobilizations, orthotics, or postural re-education.[320,323,445] Patients with chronic myofascial pain syndrome usually present with poor postures and muscle imbalances with both adaptively shortened and lengthened muscles. Strengthening shortened muscles will not correct muscular imbalances and abnormal posture, and may cause further aggravation of active myofascial trigger points, and increase pain and dysfunction. Overstretching must be avoided as this may trigger myofascial trigger points. Prior to initiating isotonic training and conditioning programs, abnormal postures must be corrected. Already during the pain-control phase of the program, patients can correct their postures and muscle imbalances by gently stretching shortened muscles, improving neural mobility, and restoring basic function. Correction and prevention of abnormal postures require a comprehensive program to include exercises to restore normal dynamic vertebral stabilization and mobility, motor control, muscle balances, strength, endurance, and breathing patterns. Many patients are aerobically deconditioned, which, combined with poor posture, may cause adaptive shortening of the auxiliary respiratory muscles, such as the scalenes, restricted chest expansions, and paradoxical breathing. Paradoxical breathing should be corrected with functional abdominal breathing.[305,323,446,447] Certain work tasks or activities of daily living may predispose a patient to chronic musculoskeletal overload, increasing the risk of myofascial dysfunction. Considering activity-related aspects of myofascial pain syndrome will enhance treatment outcomes. Modifying the workplace or the patient's work habits can be critical. If a patient continues to be exposed to certain workplace or other stress factors without modification of the conditions, the potential cause of myofascial dysfunction may not be addressed adequately. Throughout the treatment process, much attention should be paid to educating the patient regarding the etiology, perpetuating factors, and self-management. In patients with chronic myofascial pain, psychosocial issues must be assessed and addressed as outlined in the section on fibromyalgia. Patients must learn to modify their behaviors and avoid overloading the muscles without resorting to total inactivity.[309]

SOFT TISSUE LESION AND MECHANICAL DYSFUNCTION

Definition and Characteristics of Soft Tissue Mechanical Dysfunction

Fibromyalgia, with its lack of specific diagnostic findings and diffuse pain patterns, represents one end of a spectrum of pain severity and complexity and soft tissue mechanical dysfunction represents the other end. Mechanical dysfunction, where mechanical pathology exists and can be diagnosed, afflicts the greater portion of patients with acute pain. There is usually overuse or direct trauma to the tissue that causes inflammation. A partial or full tear, as in a hamstring tear or "pull," gastrocnemius tear, tennis elbow, or de Quervain's disease, for example, are forms of soft tissue mechanical dysfunction. Facet hypomobility or hypermobility, muscular or movement imbalances, discogenic pathologies, and sacroiliac joint dysfunction, for instance, all represent mechanical dysfunction characterized by soft tissue lesions. These dysfunctions can be medically diagnosed and evaluated for specific pathologies. Treatment can commence based on evaluative findings and the condition and reactivity of the tissue. Once soft tissue mechanical dysfunction becomes more subacute or chronic, clinicians should consider whether myofascial trigger points have become the main factor and, if so, alter the treatment strategy accordingly.

The specific evaluation process for soft tissue mechanical dysfunction requires a systematic approach. Looking for reproduction of pain based on palpation, muscle contraction, or stretch helps to localize the dysfunction to a specific lesion. The purpose is to identify and define areas of somatic dysfunction and to localize a lesion site. Somatic dysfunction can be defined as impaired or altered function of related components of the somatic system (body framework), skeletal, arthrodial, and myofascial structures. The criteria for dysfunction consist of:

- Structural or functional asymmetry of related parts of the musculoskeletal system, ascertained by observation and palpation.
- Tissue texture abnormality of the musculoskeletal system soft tissues (skin, fascia, muscle, ligament, or joint capsule) ascertained by observation and palpation.
- Range-of-motion abnormality of a joint, several joints, or regions of the musculoskeletal system (either restricted or hypermobile, qualitative changes in range of motion such as cogwheel movement, hesitations, and compensations) ascertained by

observations and palpation, utilizing both active and passive testing.

Management of Soft Tissue Mechanical Dysfunction

The clinical history will usually offer substantial clues to causes of the dysfunction, such as trauma, overuse, or lifestyle, among others. The evaluation will reveal specific findings that will allow for systematic development of treatment plans specific to the particular pathology or dysfunction. Treatment is usually much shorter term, and the prognosis for recovery is the best of the three categories described.

REFERENCES

1. Simons DG. Muscle pain syndromes—part 1. *Am J Phys Med*. 1975;54:289–311.
2. Stockman R. The causes, pathology, and treatment of chronic rheumatism. *Edinburgh Med J*. 1904;15:107–116.
3. Lange F, Eversbusch G. Die Bedeutung der Muskelhärten für die allgemeine Praxis. *Münch Med Wochenschr*. 1921;68:418–420.
4. Lange M. *Die Muskelhärten (Myogelosen)*. München: J.F. Lehmann's Verlag; 1931.
5. Travell JG, Rinzler S, et al. Pain and disability of the shoulder and arm. *JAMA*. 1942;120:417–422.
6. Travell JG, Rinzler SH. The myofascial genesis of pain. *Postgrad Med*. 1952;11:452–434.
7. Simons DG. Triggerpunkte und Myogelose. *Manuelle Medizin*. 1997;35:290–294.
8. Gowers WR. A lecture on lumbago: its lessons and analogues. *BMJ*. 1904;1:117–121.
9. Smythe HA, Moldofsky H. Two contributions to understanding of the "fibrositis" syndrome. *Bull Rheum Dis*. 1977;28:928–931.
10. Yunus MB. Fibromyalgia syndrome: a need for uniform classification [editorial]. *J Rheumatol*. 1983;10:841–844.
11. Wolfe F. Development of criteria for the diagnosis of fibrositis. *Am J Med*. 1986;81(3A):99–104.
12. Yunus MB, Kalyan-Raman UP, et al. Primary fibromyalgia syndrome and myofascial pain syndrome: clinical features and muscle pathology. *Arch Phys Med Rehabil*. 1988;69(6):451–454.
13. Goldenberg DL. Fibromyalgia and its relation to chronic fatigue syndrome, viral illness and immune abnormalities. *J Rheumatol Suppl*. 1989;19:91–93.
14. Berg PA, Klein R. Fibromyalgie-Syndrom; eine neuroendokrinologische Autoimmunerkrankung? *Dtsch Med Wochenschr*. 1994;119(12):429–435.
15. Wolfe F, Ross K, et al. The prevalence and characteristics of fibromyalgia in the general population. *Arthritis Rheum*. 1995;38:19–28.
16. Buskila D, Shnaider A, et al. Fibromyalgia in hepatitis C virus infection. Another infectious disease relationship. *Arch Intern Med*. 1997;157(21):2497–2500.
17. Clauw DJ, Schmidt M, et al. The relationship between fibromyalgia and interstitial cystitis. *J Psychiatr Res*. 1997;31(1):125–131.
18. Ackenheil M. Genetics and pathophysiology of affective disorders: relationship to fibromyalgia. *Z Rheumatol*. 1998;57(Suppl 2):5–7.
19. Crofford LJ. The hypothalamic–pituitary–adrenal stress axis in fibromyalgia and chronic fatigue syndrome. *Z Rheumatol*. 1998;57(Suppl 2):67–71.
20. Krause K-H, Krause J, et al. Fibromyalgia syndrome and attention deficit hyperactivity disorder: is there a comorbidity and are there consequences for the therapy of fibromyalgia syndrome? *J Musculoske Pain*. 1998;6(4):111–116.
21. Offenbaecher M, Glatzeder K, et al. Self-reported depression, familial history of depression and fibromyalgia (FM), and psychological distress in patients with FM. *Z Rheumatol*. 1998;57(Suppl 2):94–96.

22. Adler GK, Kinsley BT, et al. Reduced hypothalamic–pituitary and sympathoadrenal responses to hypoglycemia in women with fibromyalgia syndrome. *Am J Med.* 1999;106(5):534–543.

23. Buskila D, Odes LR, et al. Fibromyalgia in inflammatory bowel disease. *J Rheumatol.* 1999;26(5):1167–1171.

24. Masi AT. Review of the epidemiology and criteria of fibromyalgia and myofascial pain syndrome: concepts of illness in populations as applied to dysfunctional syndromes. *J Musculoske Pain.* 1993;1(3/4):113–136.

25. Yunus MB. Psychological aspects of fibromyalgia syndrome: a component of the dysfunctional spectrum syndrome. *Baillières Clin Rheumatol.* 1994;8(4):811–837.

26. Masi AT. Concepts of illness in populations as applied to fibromyalgia syndromes: a biopsychosocial perspective. *Z Rheumatol.* 1998;57(Suppl 2):31–35.

27. Clark P, Burgos-Vargas R, et al. Prevalence of fibromyalgia in children: a clinical study of Mexican children. *J Rheumatol.* 1998;25(10):2009–2014.

28. Wolfe F, Smythe HA, et al. The American College of Rheumatology 1990 Criteria for the Classification of Fibromyalgia. Report of the Multicenter Criteria Committee. *Arthritis Rheum.* 1990;33(2):160–172.

29. Consensus document on fibromyalgia: the Copenhagen Declaration. *J Musculoske Pain.* 1993;1(3/4):295–312.

30. Wolfe F. The future of fibromyalgia: some critical issues. *J Musculoske Pain.* 1995;3(2):3–15.

31. Bennett R. Chronic widespread pain and the fibromyalgia construct. *International Association for the Study of Pain—SIG on Rheumatic Pain Newsletter.* 1999; January:2–7.

32. McBeth J, Macfarlane GJ, et al. The association between tender points, psychological distress, and adverse childhood experiences: a community-based study. *Arthritis Rheum.* 1999;42(7):1397–1404.

33. Croft P, Burt J, et al. More pain, more tender points: is fibromyalgia just one end of a continuous spectrum? *Ann Rheum Dis.* 1996;55(7):482–485.

34. Wolfe F. The relation between tender points and fibromyalgia symptom variables: evidence that fibromyalgia is not a discrete disorder in the clinic. *Ann Rheum Dis.* 1997;56:268–271.

35. Turk DC, Okifuji A. Assessment of patients' reporting of pain: an integrated perspective. *Lancet.* 1999;353(9166):1784–1788.

36. Jacobs JW, Rasker JJ, et al. Lack of correlation between the mean tender point score and self-reported pain in fibromyalgia. *Arthritis Care Res.* 1996;9(2):105–111.

37. Kuncl RW, George EB. Toxic neuropathies and myopathies. *Curr Opin Neurol.* 1993;6(5):695–704.

38. Marin R, Connick E. Tension myalgia versus myoadenylate deaminase deficiency: a case report. *Arch Phys Med Rehabil.* 1997;78(1):95–97.

39. Gerwin RD. Myofascial pain and fibromyalgia: diagnosis and treatment. *J Back and Musculoskeletal Rehab.* 1998;11:175–181.

40. Menninger H. Other pain syndromes to be differentiated from fibromyalgia. *Z Rheumatol.* 1998;57(Suppl 2):56–60.

41. Wierzbicki AS, Lumb PJ, et al. High-dose atorvastatin therapy in severe heterozygous familial hypercholesterolaemia. *QJM.* 1998;91(4):291–294.

42. Wolfe F. Fibromyalgia: on diagnosis and certainty. *J Musculoske Pain.* 1993;1(3/4):17–35.

43. Wolfe F. The fibromyalgia problem [editorial]. *J Rheumatol.* 1997;24(7):1247–1249.

44. Toombs SK. *The Meaning of Illness: A Phenomenological Account of the Different Perspectives of Physician and Patient.* Dordrecht: Kluwer Academic Publishers; 1992.

45. Annells M. Hermeneutic phenomenology: philosophical perspectives and current use in nursing research. *J Adv Nurs.* 1996;23(4):705–713.

46. Hadler NM. If you have to prove you are ill, you can't get well. The object lesson of fibromyalgia. *Spine.* 1996; 21(20):2397–2400.

47. Solomon DH, Liang MH. Fibromyalgia: scourge of humankind or bane of a rheumatologist's existence? [editorial]. *Arthritis Rheum.* 1997;40(9):1553–1555.

48. Wolfe F, Anderson J, et al. Health status and disease severity in fibromyalgia. *Arthritis Rheum.* 1997;40(9):1571–1579.

49. Hadler NM. Fibromyalgia, chronic fatigue, and other iatrogenic diagnostic algorithms. Do some labels escalate illness in vulnerable patients? *Postgrad Med.* 1997;102(2):161–172.

50. DeVellis BM, Blalock SJ. Illness attributions and hopelessness depression: the role of hopelessness expectancy. *J Abnorm Psychol.* 1992;101(2):257–264.

51. Jensen MP, Romano JM, et al. Patient beliefs predict patient functioning: further support for a cognitive-behavioural model of chronic pain. *Pain.* 1999; 81(1–2):95–104.

52. Hellström O, Bullington J, et al. A phenomenological study of fibromyalgia. Patient perspectives. *Scand J Prim Health Care.* 1999;17(1):11–16.

53. Aaron LA, Bradley LA, et al. Psychiatric diagnoses in patients with fibromyalgia are related to health care-seeking behavior rather than to illness. *Arthritis Rheum.* 1996;39(3):436–445.

54. Russell IJ. Fibromyalgia syndrome: approaches to management. *Bull Rheum Dis.* 1996;45(3):1–4.

55. Wolfe F, Hawley DJ. Evidence of disordered symptom appraisal in fibromyalgia: increased rates of reported comorbidity and comorbidity severity. *Clin Exp Rheumatol.* 1999;17(3):297–303.

56. Poduri KR, Gibson CJ. Drug related lupus misdiagnosed as fibromyalgia: case report. *J Musculoske Pain.* 1995;3(4):71–78.

57. Müller B, Müller W. Die generalisierte Tendomyopathie (Fibromyalgie). *Z Gesamte Inn Med.* 1991; 46(10–11):361–369.

58. Samborski W, Stratz T, et al. Druckpunktuntersuchungen bei der generalisierten Tendomyopathie (Fibromyalgie) (Vergleich verschiedener Methoden). *Z Rheumatol.* 1991;50(6):382–386.

59. Häntzschel H, Boche K. Das Fibromyalgiasyndrom. *Fortschr Med.* 1999;117(5):26–31.

60. Goldenberg DL. Fibromyalgia syndrome a decade later: what have we learned? *Arch Intern Med.* 1999; 159(8):777–785.

61. Cohen ML, Quintner JL. Fibromyalgia syndrome, a problem of tautology. *Lancet.* 1993;342(8876):906–909.

62. Cohen ML, Quintner JL. Fibromyalgia syndrome and disability: a failed construct fails those in pain. *Med J Aust.* 1998;168(8):402–404.

63. Quintner JL, Cohen ML. Fibromyalgia falls foul of a fallacy. *Lancet.* 1999;353(9158):1092–1094.

64. Lautenbacher S, Rollman GB, et al. Multi-method assessment of experimental and clinical pain in patients with fibromyalgia. *Pain.* 1994;59(1):45–53.

65. Kosek E, Ekholm J, et al. Increased pressure pain sensibility in fibromyalgia patients is located deep to the skin but not restricted to muscle tissue. *Pain.* 1995; 63(3):335–339.

66. Vecchiet L, Pizzigallo E, et al. Differentiation of sensitivity in different tissues and its clinical significance. *J Musculoske Pain.* 1998;6(1):33–45.

67. Granges G, Littlejohn G. Prevalence of myofascial pain syndrome in fibromyalgia syndrome and regional pain syndrome: a comparative study. *J Musculoske Pain.* 1993;1(2):19–36.

68. Russell IJ. The reliability of algometry in the assessment of patients with fibromyalgia syndrome. *J Musculoske Pain.* 1998;6(1):139–152.

69. Tunks E, Crook J, et al. Tender points in fibromyalgia. *Pain.* 1988;34(1):11–19.

70. Fischer AA. New developments in diagnosis of myofascial pain and fibromyalgia. In: Fischer AA, ed. *Myofascial Pain: Update in Diagnosis and Treatment.* Philadelphia: WB Saunders; 1997:1–21.

71. Okifuji A, Turk DC, et al. A standardized manual tender point survey. I. Development and determination of a threshold point for the identification of positive tender points in fibromyalgia syndrome. *J Rheumatol.* 1997; 24(2):377–383.

72. Tunks E, McCain GA, et al. The reliability of examination for tenderness in patients with myofascial pain,

chronic fibromyalgia and controls. *J Rheumatol.* 1995; 22(5):944–952.

73. Fischer AA. Algometry in diagnosis of musculoskeletal pain and evaluation of treatment outcome: an update. *J Musculoske Pain.* 1998;6(1):5–32.

74. Puttick M, Schulzer M, et al. Reliability and reproducibility of fibromyalgic tenderness, measurement by electronic and mechanical dolorimeters. *J Musculoske Pain.* 1995;3(4):3–14.

75. Zohn DA, Clauw DJ. A comparison of skin rolling and tender points as a diagnostic test for fibromyalgia. *J Musculoske Pain.* 1999;7(3):127–136.

76. Cohen ML, Sheather-Reid RB, et al. Evidence for abnormal nociception in fibromyalgia and repetitive strain injury. *J Musculoke Pain.* 1995;3(2):49–57.

77. McDermid AJ, Rollman GB, et al. Generalized hypervigilance in fibromyalgia: evidence of perceptual amplification. *Pain.* 1996;66(2–3):133–144.

78. Bendtsen L, Nørregaard J, et al. Evidence of qualitatively altered nociception in patients with fibromyalgia. *Arthritis Rheum.* 1997;40(1):98–102.

79. Lorenz J. Hyperalgesia or hypervigilance? An evoked potential approach to the study of fibromyalgia syndrome. *Z Rheumatol.* 1998;57(Suppl 2):19–22.

80. Kosek E, Ekholm J, et al. Modulation of pressure pain thresholds during and following isometric contraction in patients with fibromyalgia and in healthy controls. *Pain.* 1996;64(3):415–423.

81. Gibson SJ, Littlejohn GO, et al. Altered heat pain thresholds and cerebral event-related potentials following painful CO_2 laser stimulation in subjects with fibromyalgia syndrome. *Pain.* 1994;58(2):185–193.

82. Nørregaard J, Bendtsen L, et al. Pressure and heat pain thresholds and tolerances in patients with fibromyalgia. *J Musculoske Pain.* 1996;5(2):43–53.

83. Yunus MB, Kalyan-Raman UP, et al. Electron microscopic studies of muscle biopsy in primary fibromyalgia syndrome: a controlled and blinded study. *J Rheumatol.* 1989;16(1):97–101.

84. Wolfe F. Fibromyalgia and problems in classification of musculoskeletal disorders. In: Værøy H, Merskey H, eds. *Progress in Fibromyalgia and Myofascial Pain.* Amsterdam: Elsevier; 1993:217–235.

85. Schrøder HD, Drewes AM, et al. Muscle biopsy in fibromyalgia. *J Musculoske Pain.* 1993;1(3/4):165–169.

86. Graven-Nielsen T, Sörensen J, et al. Central hyperexcitability in fibromyalgia. *J Musculoske Pain.* 1999; 7(1/2):261–271.

87. Mountz JM, Bradley LA, et al. Fibromyalgia in women. Abnormalities of regional cerebral blood flow in the thalamus and the caudate nucleus are associated with low pain threshold levels. *Arthritis Rheum.* 1995;38(7): 926–938.

88. Bradley LA, Sotolongo A, et al. Abnormal regional cerebral blood flow in the caudate nucleus among fibromyalgia patients and non-patients is associated with insidious symptom onset. *J Musculoske Pain.* 1999;7(1/2):285–292.

89. Clauw DJ. Fibromyalgia: more than just a musculoskeletal disease. *Am Fam Physician.* 1995;52(3):843–854.

90. White KP, Speechley M, et al. Comparing self-reported function and work disability in 100 community cases of fibromyalgia syndrome versus controls in London, Ontario: the London Fibromyalgia Epidemiology Study. *Arthritis Rheum.* 1999;42(1):76–83.

91. Hauri P, Hawkins DR. Alpha-delta sleep. *Electroencephalogr Clin Neurophysiol.* 1973;34(3):233–237.

92. Oswald I. Sleep as restorative process: human clues. *Prog Brain Res.* 1980;53:279–288.

93. Moldofsky H, Scarisbrick P, et al. Musculosketal symptoms and non-REM sleep disturbance in patients with "fibrositis syndrome" and healthy subjects. *Psychosom Med.* 1975;37(4):341–351.

94. Molony RR, MacPeek DM, et al. Sleep, sleep apnea and the fibromyalgia syndrome. *J Rheumatol.* 1986; 13(4):797–800.

95. Drewes AM, Nielsen KD, et al. Alpha intrusion in fibromyalgia. *J Musculoske Pain.* 1993;1(3/4):223–228.

96. Branco J, Atalaia A, et al. Sleep cycles and alpha-delta sleep in fibromyalgia syndrome. *J Rheumatol.* 1994; 21(6):1113–1117.

97. Drewes AM, Gade K, et al. Clustering of sleep electroencephalographic patterns in patients with the fibromyalgia syndrome. *Br J Rheumatol.* 1995;34(12): 1151–1156.

98. Older SA, Battafarano DF, et al. The effects of delta wave sleep interruption on pain thresholds and fibromyalgia-like symptoms in healthy subjects; correlations with insulin-like growth factor I. *J Rheumatol.* 1998;25(6):1180–1186.

99. Carette S, Oakson G, et al. Sleep electroencephalography and the clinical response to amitriptyline in patients with fibromyalgia. *Arthritis Rheum.* 1995; 38(9):1211–1217.

100. Scheuler W, Stinshoff D, et al. The alpha-sleep pattern. Differentiation from other sleep patterns and effect of hypnotics. *Neuropsychobiology.* 1983;10(2–3):183–189.

101. Moldofsky H, Lue FA, et al. Sleep and morning pain in primary osteoarthritis. *J Rheumatol.* 1987;14(1): 124–128.

102. Kubicki S, Henkes H, et al. Schlafpolygraphische Daten von AIDS-Patienten. *EEG EMG Z Elektroenzephalogr Elektromyogr Verwandte Geb.* 1989;20(4):288–294.

103. Hirsch M, Carlander B, et al. Objective and subjective sleep disturbances in patients with rheumatoid arthritis. A reappraisal. *Arthritis Rheum.* 1994;37(1):41–49.

104. Scudds RA, Trachsel LC, et al. A comparative study of pain, sleep quality and pain responsiveness in fibrositis and myofascial pain syndrome. *J Rheumatol.* 1989;16(19) (Suppl):120–126.

105. Lue FA. Sleep and fibromyalgia. *J Musculoske Pain.* 1994;2(3):89–100.

106. Hudson JI, Pope HG. The concept of affective spectrum disorder: relationship to fibromyalgia and other syndromes of chronic fatigue and chronic muscle pain. *Baillières Clin Rheumatol.* 1994;8(4):839–856.

107. Yunus MB. Psychological factors in fibromyalgia syndrome. *J Musculoske Pain.* 1994;2(1):87–91.

108. Walter B, Vaitl D, et al. Affective distress in fibromyalgia syndrome is associated with pain severity. *Z Rheumatol.* 1998;57(Suppl 2):101–104.

109. Clark S, Campbell SM, et al. Clinical characteristics of fibrositis. II. A "blinded," controlled study using standard psychological tests. *Arthritis Rheum.* 1985; 28(2):132–137.

110. Mau W, Danz-Neeff H, et al. Typ-A-Verhalten und Kontrollambitionen bei Patienten mit einem primären fibromyalgischen Syndrom. In: Müller W, ed. *Generalisierte Tendomyopathie (Fibromyalgie).* Darmstadt: Steinkopff; 1991:211–213.

111. Keel P. Psychological and psychiatric aspects of fibromyalgia syndrome (FMS). *Z Rheumatol.* 1998;57(Suppl 2):97–100.

112. Blumer D, Heilbronn M. The pain-prone disorder: a clinical and psychological profile. *Psychosomatics.* 1981;22(5):395–397, 401–402.

113. Boisset-Pioro MH, Esdaile JM, et al. Sexual and physical abuse in women with fibromyalgia syndrome. *Arthritis Rheum.* 1995;38(2):235–241.

114. Taylor ML, Trotter DR, et al. The prevalence of sexual abuse in women with fibromyalgia. *Arthritis Rheum.* 1995;38(2):229–234.

115. Katz RS, Kravitz HM. Fibromyalgia, depression, and alcoholism: a family history study. *J Rheumatol.* 1996; 23(1):149–154.

116. Amir M, Kaplan Z, et al. Posttraumatic stress disorder, tenderness and fibromyalgia. *J Psychosom Res.* 1997;42(6):607–613.

117. Goldberg RT, Pachas WN, et al. Relationship between traumatic events in childhood and chronic pain. *Disabil Rehabil.* 1999;21(1):23–30.

118. Piergiacomi G, Blasetti P, et al. Personality pattern in rheumatoid arthritis and fibromyalgic syndrome. Psychological investigation. *Z Rheumatol.* 1989;48(6): 288–293.

119. Ahles TA, Khan SA, et al. Psychiatric status of patients with primary fibromyalgia, patients with rheu-

matoid arthritis, and subjects without pain: a blind comparison of DSM-III diagnoses. *Am J Psychiatry.* 1991;148(12):1721–1726.

120. Yunus MB, Ahles TA, et al. Relationship of clinical features with psychological status in primary fibromyalgia. *Arthritis Rheum.* 1991;34(1):15–21.

121. Hudson JI, Pope, HG, Jr. The relationship between fibromyalgia and major depressive disorder. *Rheum Dis Clin North Am.* 1996;22(2):285–303.

122. Neeck G, Riedel W. Hormonal pertubations in fibromyalgia syndrome. *Ann N Y Acad Sci.* 1999;876: 325–338.

123. Fassbender K, Samborsky W, et al. Tender points, depressive and functional symptoms: comparison between fibromyalgia and major depression. *Clin Rheumatol.* 1997;16(1):76–79.

124. Burckhardt CS, Clark SR, et al. Fibromyalgia and quality of life: a comparative analysis. *J Rheumatol.* 1993;20(3):475–479.

125. Hawley DJ, Wolfe F. Depression is not more common in rheumatoid arthritis: a 10-year longitudinal study of 6,153 patients with rheumatic disease. *J Rheumatol.* 1993;20(12):2025–2031.

126. Walker EA, Keegan D, et al. Psychosocial factors in fibromyalgia compared with rheumatoid arthritis: I. Psychiatric diagnoses and functional disability. *Psychosom Med.* 1997;59(6):565–571.

127. Bohr TW. Fibromyalgia syndrome and myofascial pain syndrome. Do they exist? *Neurol Clin.* 1995;13(2): 365–384.

128. Ford CV. Somatization and fashionable diagnoses: illness as a way of life. *Scand J Work Environ Health.* 1997;23(Suppl 3):7–16.

129. Haynes RB, Sackett DL, et al. Increased absenteeism from work after detection and labeling of hypertensive patients. *N Engl J Med.* 1978;299(14):741–744.

130. Wolfe F, Anderson J, et al. Work and disability status of persons with fibromyalgia. *J Rheumatol.* 1997; 24(6):1171–1178.

131. Barsky AJ, Borus JF. Functional somatic syndromes. *Ann Intern Med.* 1999;130:910–921.

132. Elks ML. On the genesis of somatization disorder: the role of the medical profession. *Med Hypotheses.* 1994;43(3):151–154.

133. Kouyanou K, Pither CE, et al. Iatrogenic factors and chronic pain. *Psychosom Med.* 1997;59:597–604.

134. Stewart DE. The changing faces of somatization. *Psychosomatics.* 1990;31:153–158.

135. Shorter E. *From Paralysis to Fatigue: A History of Psychosomatic Illness in the Modern Era.* New York: Free Press; 1992.

136. Carette S. Fibromyalgia 20 years later: what have we

really accomplished? [editorial]. *J Rheumatol.* 1995; 22(4):590–594.

137. Mannerkorpi K, Burckhardt CS, et al. Physical performance characteristics of women with fibromyalgia. *Arthritis Care Res.* 1994;7(3):123–129.

138. Mengshoel AM. Fibromyalgia and responses to exercise. *J Manual Manipulative Therapy.* 1998;6(3): 144–150.

139. Sietsema KE, Cooper DM, et al. Oxygen uptake during exercise in patients with primary fibromyalgia syndrome. *J Rheumatol.* 1993;20(5):860–865.

140. Vestergaard-Poulsen P, Thomsen C, et al. 31P NMR spectroscopy and electromyography during exercise and recovery in patients with fibromyalgia. *J Rheumatol.* 1995;22(8):1544–1551.

141. Mengshoel AM, Saugen E, et al. Muscle fatigue in early fibromyalgia. *J Rheumatol.* 1995;22(1):143–150.

142. Nørregaard J, Bulow PM, et al. Biochemical changes in relation to a maximal exercise test in patients with fibromyalgia. *Clin Physiol.* 1994;14(2):159–167.

143. Jubrias SA, Bennett RM, et al. Increased incidence of a resonance in the phosphodiester region of 31P nuclear magnetic resonance spectra in the skeletal muscle of fibromyalgia patients. *Arthritis Rheum.* 1994;37(6): 801–807.

144. Mengshoel A, Vøllestad N, et al. Pain and fatigue induced by exercise in fibromyalgia patients and sedentary healthy subjects. *Clin Exp Rheumatol.* 1995; 13:477–482.

145. Zimmermann M. Pathophysiological mechanisms of fibromyalgia. *Clin J Pain.* 1991;7(Suppl 1):S8–15.

146. Elam M, Johansson G, et al. Do patients with primary fibromyalgia have an altered muscle sympathetic nerve activity? *Pain.* 1992;48(3):371–375.

147. Henriksson KG, Bengtsson A, et al. Muscle biopsy findings of possible diagnostic importance in primary fibromyalgia (fibrositis, myofascial syndrome). *Lancet.* 1982;2(8312):1395.

148. Bartels EM, Danneskiold-Samsøe B. Histological abnormalities in muscle from patients with certain types of fibrositis. *Lancet.* 1986;1(8484):755–757.

149. Bengtsson A, Henriksson KG, et al. Muscle biopsy in primary fibromyalgia. Light-microscopical and histochemical findings. *Scand J Rheumatol.* 1986; 15(1):1–6.

150. Lund N, Bengtsson A, et al. Muscle tissue oxygen pressure in primary fibromyalgia. *Scand J Rheumatol.* 1986;15(2):165–173.

151. Larsson SE, Bengtsson A, et al. Muscle changes in work-related chronic myalgia. *Acta Orthop Scand.* 1988;59(5):552–556.

152. Bengtsson A, Henriksson KG. The muscle in fibro-

myalgia—a review of Swedish studies. *J Rheumatol Suppl.* 1989;19:144–149.

153. Jacobsen S, Bartels EM, et al. Single cell morphology of muscle in patients with chronic muscle pain. *Scand J Rheumatol.* 1991;20(5):336–343.

154. Henriksson KG, Bengtsson A, et al. Morphological changes in muscle in fibromyalgia and chronic shoulder myalgia. In: Værøy H, Merskey H, eds. *Progress in Fibromyalgia and Myofascial Pain.* Amsterdam: Elsevier; 1993:61–73.

155. Sprott H, Muller A, et al. Collagen crosslinks in fibromyalgia. *Arthritis Rheum.* 1997;40(8):1450–1454.

156. Nørregaard J, Harreby M, et al. Single cell morphology and high-energy phosphate levels in quadriceps muscles from patients with fibromyalgia. *J Musculoske Pain.* 1994;2(2):45–51.

157. Pongratz DE, Späth M. Morphologic aspects of fibromyalgia. *Z Rheumatol.* 1998;57(Suppl 2):47–51.

158. Simms RW, Roy SH, et al. Lack of association between fibromyalgia syndrome and abnormalities in muscle energy metabolism. *Arthritis Rheum.* 1994; 37(6):794–800.

159. Simms RW. Muscle studies in fibromyalgia. *J Musculoske Pain.* 1994;2(3):117–123.

160. Kravis MM, Munk PL, et al. MR imaging of muscle and tender points in fibromyalgia. *J Magn Reson Imaging.* 1993;3(4):669–670.

161. Simms RW. Is there muscle pathology in fibromyalgia syndrome? *Rheum Dis Clin North Am.* 1996;22(2): 245–266.

162. Gifford LS, Butler DS. The integration of pain sciences into clinical practice. *J Hand Ther.* 1997;10:86–95.

163. Værøy H, Helle R, et al. Elevated CSF levels of substance P and high incidence of Raynaud phenomenon in patients with fibromyalgia: new features for diagnosis. *Pain.* 1988;32(1):21–26.

164. Russell IJ, Orr MD, et al. Elevated cerebrospinal fluid levels of substance P in patients with the fibromyalgia syndrome. *Arthritis Rheum.* 1994;37(11):1593–1601.

165. Welin M, Bragee B, et al. Elevated substance P levels are contrasted by a decrease in met-enkephalin-arg-phe levels in CSF from fibromyalgia patients (abstract). *J Musculoske Pain.* 1995;3(Suppl 1):4.

166. Almay BG, Johansson F, et al. Substance P in CSF of patients with chronic pain syndromes. *Pain.* 1988; 33(1):3–9.

167. Cramer H, Rosler N, et al. Cerebrospinal fluid immunoreactive substance P and somatostatin in neurological patients with peripheral and spinal cord disease. *Neuropeptides.* 1988;12(3):119–124.

168. Nyberg F, Liu Z, et al. Enhanced CSF levels of substance P in patients with painful arthrosis but not in patients with pain from herniated lumbar discs (abstract). *J Musculoske Pain.* 1995;3(Suppl 1):2.

169. Sprott H, Bradley LA, et al. Immunohistochemical and molecular studies of serotonin, substance P, galanin, pituitary adenylyl cyclase-activating polypeptide and secretoneurin in fibromyalgic muscle tissue. *Arthritis Rheum.* 1998;41(9):1689–1694.

170. Doyle C, Palmer JA, et al. Molecular consequences of noxious stimulation. In: Borsook D, ed. *Molecular Neurobiology of Pain.* Seattle: IASP Press; 1997: 145–169.

171. Rygh LJ, Svendsen F, et al. Natural noxious stimulation can induce long-term increase of spinal nociceptive responses. *Pain.* 1999;82(3):305–310.

172. Coderre TJ, Katz J, et al. Contribution of central neuroplasticity to pathological pain: review of clinical and experimental evidence. *Pain.* 1993;52(3):259–285.

173. Hoheisel U, Mense S, et al. Effects of spinal cord superfusion with substance P on the excitability of rat dorsal horn neurons processing input from deep tissues. *J Musculoske Pain.* 1996;3(3):23–43.

174. Yaksh TL, Hua XY, et al. The spinal biology in humans and animals of pain states generated by persistent small afferent input. *Proc Natl Acad Sci U S A.* 1999; 96(14):7680–7686.

175. Maggi CA, Patacchini R, et al. Tachykinin receptors and tachykinin receptor antagonists. *J Auton Pharmacol.* 1993;13(1):23–93.

176. Giovengo SL, Russell IJ, et al. Increased concentrations of nerve growth factor in cerebrospinal fluid of patients with fibromyalgia. *J Rheumatol.* 1999; 26(7):1564–1569.

177. Russell IJ. Neurochemical pathogenesis of fibromyalgia syndrome. *J Musculoske Pain.* 1999;7(1/2):183–191.

178. Heils A, Mossner R, et al. The human serotonin transporter gene polymorphism—basic research and clinical implications. *J Neural Transm.* 1997;104(10): 1005–1014.

179. Lefebvre H, Contesse V, et al. The 5-HT receptor in the adrenal gland. In: Eglen RM, ed. *5-HT4 receptors in the brain and periphery.* Berlin: Springer; 1998:195–211.

180. Sapolsky RM. Neuroendocrinology of the stress-response. In: Becker JB, Breedlove SM, et al, eds. *Behavioral Endocrinology.* Cambridge: MIT Press; 1992: 287–324.

181. Crofford LJ, Demitrack MA. Evidence that abnormalities of central neurohormonal systems are key to understanding fibromyalgia and chronic fatigue syndrome. *Rheum Dis Clin North Am.* 1996;22(2):267–284.

182. Aron DC, Findling JW, et al. Hypothalamus and pituitary. In: Greenspan FS, Strewler GJ, eds. *Basic and Clinical Endocrinology.* Stamford: Appleton & Lange; 1997:95–156.

183. Hegde SS. 5-HT4 receptors in gastrointestinal tract. In: Eglen RM, ed. *5-HT4 Receptors in the Brain and Periphery*. Berlin: Springer; 1998:149–169.

184. Harvey JA, Schlosberg AJ, et al. Behavioral correlates of serotonin depletion. *Fed Proc*. 1975;34(9): 1796–1801.

185. Moldofsky H, Warsh JJ. Plasma tryptophan and musculoskeletal pain in non-articular rheumatism ("fibrositis syndrome"). *Pain*. 1978;5(1):65–71.

186. Russell IJ. Substance P and fibromyalgia. *J Musculoske Pain*. 1998;6(3):29–35.

187. Russell IJ, Værøy H, et al. Cerebrospinal fluid biogenic amine metabolites in fibromyalgia/fibrositis syndrome and rheumatoid arthritis. *Arthritis Rheum*. 1992;35(5):550–556.

188. Russell IJ, Vipraio GA, et al. Abnormalities in the central nervous system (CNS) metabolism of tryptophan (TRY) to 3-hydroxy kynurenine (OHKY) in fibromyalgia syndrome (FS) (abstract). *Arthritis Rheum*. 1993;36(9):S222.

189. Schwarz MJ, Spath M, et al. Relationship of substance P, 5-hydroxyindole acetic acid and tryptophan in serum of fibromyalgia patients. *Neurosci Lett*. 1999; 259(3):196–198.

190. Hrycaj P, Stratz T, et al. Platelet 3H-imipramine uptake receptor density and serum serotonin levels in patients with fibromyalgia/fibrositis syndrome. *J Rheumatol*. 1993;20(11):1986–1988.

191. Stratz T, Samborski W, et al. Die Serotoninkonzentration im Serum bei Patienten mit generalisierter Tendomyopathie (Fibromyalgie) und chronischer Polyarthritis. *Med Klin*. 1993;88(8):458–462.

192. Russell IJ. Biochemical abnormalities in fibromyalgia syndrome. *J Musculoske Pain*. 1994;2(3):101–115.

193. Wolfe F, Russell IJ, et al. Serotonin levels, pain threshold, and fibromyalgia symptoms in the general population. *J Rheumatol*. 1997;24(3):555–559.

194. Ernberg M, Hedenberg-Magnusson B, et al. The level of serotonin in the superficial masseter muscle in relation to local pain and allodynia. *Life Sci*. 1999; 65(3):313–325.

195. Klein R, Bansch M, et al. Clinical relevance of antibodies against serotonin and gangliosides in patients with primary fibromyalgia syndrome. *Psychoneuroendocrinology*. 1992;17(6):593–598.

196. Klein R, Berg PA. High incidence of antibodies to 5-hydroxytryptamine, gangliosides and phospholipids in patients with chronic fatigue and fibromyalgia syndrome and their relatives: evidence for a clinical entity of both disorders. *Eur J Med Res*. 1995;1(1):21–26.

197. Olin R, Klein R, et al. A randomised double-blind 16-week study of ritanserin in fibromyalgia syndrome: clinical outcome and analysis of autoantibodies to serotonin, gangliosides and phospholipids. *Clin Rheumatol*. 1998;17(2):89–94.

198. Coplan JD, Tamir H, et al. Plasma anti-serotonin and serotonin anti-idiotypic antibodies are elevated in panic disorder. *Neuropsychopharmacology*. 1999;20(4): 386–391.

199. Crofford LJ, Casey KL. Central modulation of pain perception. *Rheum Dis Clin North Am*. 1999;25(1):1–13.

200. Eide PK, Hole K. Interactions between serotonin and substance P in the spinal regulation of nociception. *Brain Res*. 1991;550:225–230.

201. Mense S. Descending antinociception and fibromyalgia. *Z Rheumatol*. 1998;57(Suppl 2):23–26.

202. McCain GA, Tilbe KS. Diurnal hormone variation in fibromyalgia syndrome: a comparison with rheumatoid arthritis. *J Rheumatol Suppl*. 1989;19:154–157.

203. Crofford LJ, Pillemer SR, et al. Hypothalamic–pituitary–adrenal axis perturbations in patients with fibromyalgia. *Arthritis Rheum*. 1994;37(11):1583–1592.

204. Griep EN, Boersma JW, et al. Altered reactivity of the hypothalamic–pituitary–adrenal axis in the primary fibromyalgia syndrome. *J Rheumatol*. 1993; 20(3):469–474.

205. Riedel W, Layka H, et al. Secretory pattern of GH, TSH, thyroid hormones, ACTH, cortisol, FSH, and LH in patients with fibromyalgia syndrome following systemic injection of the relevant hypothalamic-releasing hormones. *Z Rheumatol*. 1998;57(Suppl 2):81–87.

206. Chrousos GP, Gold PW. The concepts of stress and stress system disorders. Overview of physical and behavioral homeostasis. *JAMA*. 1992;267(9):1244–1252.

207. Gallagher TF, Yoshida K, et al. ACTH and cortisol secretory patterns in man. *J Clin Endocrinol Metab*. 1973;36(6):1058–1068.

208. Griep EN, Boersma JW, et al. Function of the hypothalamic–pituitary–adrenal axis in patients with fibromyalgia and low back pain. *J Rheumatol*. 1998; 25(7):1374–1381.

209. de Goeij DC, Kvetnansky R, et al. Repeated stress-induced activation of corticotropin-releasing factor neurons enhances vasopressin stores and colocalization with corticotropin-releasing factor in the median eminence of rats. *Neuroendocrinology*. 1991; 53(2):150–159.

210. Scaccianoce S, Muscolo LA, et al. Evidence for a specific role of vasopressin in sustaining pituitary–adrenocortical stress response in the rat. *Endocrinology*. 1991;128(3138–3143.

211. Lentjes EG, Griep EN, et al. Glucocorticoid receptors, fibromyalgia and low back pain. *Psychoneuroendocrinology*. 1997;22(8):603–614.

212. Dessein PH, Shipton EA, et al. Hyposecretion of adrenal androgens and the relation of serum adrenal

steroids, serotonin and insulin-like growth factor-1 to clinical features in women with fibromyalgia. *Pain.* 1999;83(2):313–319.

213. Amandusson A, Hallbeck M, et al. Estrogen-induced alterations of spinal cord enkephalin gene expression. *Pain.* 1999;83(2):243–248.

214. Smith R, Wickings EJ, et al. Corticotropin-releasing hormone in chimpanzee and gorilla pregnancies. *J Clin Endocrinol Metab.* 1999;84(8):2820–2825.

215. Masi AT, Da Silva JA, et al. Perturbations of hypothalamic–pituitary–gonadal (HPG) axis and adrenal androgen (AA) functions in rheumatoid arthritis. *Baillières Clin Rheumatol.* 1996;10(2):295–332.

216. Parker CR, Jr. Dehydroepiandrosterone and dehydroepiandrosterone sulfate production in the human adrenal during development and aging. *Steroids.* 1999; 64(9):640–647.

217. Bonnefoy M, Kostka T, et al. Physical activity and dehydroepiandrosterone sulphate, insulin-like growth factor I and testosterone in healthy active elderly people. *Age Ageing.* 1998;27(6):745–751.

218. Sasaki T, Iwasaki K, et al. Association of working hours with biological indices related to the cardiovascular system among engineers in a machinery manufacturing company. *Ind Health.* 1999;37(4):457–463.

219. Maccario M, Mazza E, et al. Relationships between dehydroepiandrosterone-sulphate and anthropometric, metabolic and hormonal variables in a large cohort of obese women. *Clin Endocrinol (Oxf).* 1999;50(5): 595–600.

220. Bennett RM, Clark SR, et al. Low levels of somatomedin C in patients with the fibromyalgia syndrome. A possible link between sleep and muscle pain. *Arthritis Rheum.* 1992;35(10):1113–1116.

221. Griep EN, Boersma JW, et al. Pituitary release of growth hormone and prolactin in the primary fibromyalgia syndrome. *J Rheumatol.* 1994;21(11):2125–2130.

222. Bennett RM, Cook DM, et al. Hypothalamic–pituitary–insulin-like growth factor-I axis dysfunction in patients with fibromyalgia. *J Rheumatol.* 1997;24(7): 1384–1389.

223. Bennett RM. Disordered growth hormone secretion in fibromyalgia: a review of recent findings and a hypothesized etiology. *Z Rheumatol.* 1998;57(Suppl 2):72–76.

224. Leal-Cerro A, Povedano J, et al. The growth hormone (GH)-releasing hormone-GH-insulin-like growth factor-1 axis in patients with fibromyalgia syndrome. *J Clin Endocrinol Metab.* 1999;84(9):3378–3381.

225. Van Cauter E, Plat L. Physiology of growth hormone secretion during sleep. *J Pediatr.* 1996;128(5 Pt 2): S32–37.

226. Wehrenberg WB, Janowski BA, et al. Glucocorticoids:

potent inhibitors and stimulators of growth hormone secretion. *Endocrinology.* 1990;126(6):3200–3203.

227. Giustina A, Wehrenberg WB. Influence of thyroid hormones on the regulation of growth hormone secretion. *Eur J Endocrinol.* 1995;133(6):646–653.

228. Giustina A, Veldhuis JD. Pathophysiology of the neuroregulation of growth hormone secretion in experimental animals and the human. *Endocr Rev.* 1998; 19(6):717–797.

229. Nørregaard J, Bülow PM, et al. Somatomedin-C and procollagen aminoterminal peptide in fibromyalgia. *J Musculoske Pain.* 1995;3(4):33–40.

230. Nelson ME, Meredith CN, et al. Hormone and bone mineral status in endurance-trained and sedentary postmenopausal women. *J Clin Endocrinol Metab.* 1988; 66(5):927–933.

231. Yue SK. Relaxin: its role in the pathogenesis of fibromyalgia. In: Gerwin RD, Yue SK, eds. *Conference Proceedings of Advanced Topics in Myofascial Pain: Diagnosis and Treatment. Minneapolis 1999.* Minneapolis: Bethesda Health Clinic; 1999.

232. Ostensen M, Rugelsjoen A, et al. The effect of reproductive events and alterations of sex hormone levels on the symptoms of fibromyalgia. *Scand J Rheumatol.* 1997;26(5):355–360.

233. Bani D. Relaxin: a pleiotropic hormone. *Gen Pharmacol.* 1997;28(1):13–22.

234. Weiss G. Relaxin in the male. *Biol Reprod.* 1989; 40(2):197–200.

235. Goldstein, JA. Fibromyalgia syndrome: a pain modulation disorder related to altered limbic function? *Baillières Clin Rheumatol.* 1994;8(4):777–800.

236. Bani D, Failli P, et al. Relaxin activates the L-arginine–nitric oxide pathway in vascular smooth muscle cells in culture. *Hypertension.* 1998;31(6):1240–1247.

237. Mense S, Hoheisel U. New developments in the understanding of the pathophysiology of muscle pain. *J Musculoske Pain.* 1999;7(1/2):13–24.

238. Osheroff PL, Phillips HS. Autoradiographic localization of relaxin binding sites in rat brain. *Proc Natl Acad Sci U S A.* 1991;88(15):6413–6417.

239. Osheroff PL, Ho WH. Expression of relaxin mRNA and relaxin receptors in postnatal and adult rat brains and hearts. Localization and developmental patterns. *J Biol Chem.* 1993;268(20):15193–15199.

240. Parry LJ, Summerlee AJ. Central angiotensin partially mediates the pressor action of relaxin in anesthetized rats. *Endocrinology.* 1991;129(1):47–52.

241. Geddes BJ, Parry LJ, et al. Brain angiotensin-II partially mediates the effects of relaxin on vasopressin and oxytocin release in anesthetized rats. *Endocrinology.* 1994;134(3):1188–1192.

242. Parry LJ, Poterski RS, et al. Effects of relaxin on blood

pressure and the release of vasopressin and oxytocin in anesthetized rats during pregnancy and lactation. *Biol Reprod.* 1994;50(3):622–628.

243. Bethea CL, Cronin MJ, et al. The effect of relaxin infusion on prolactin and growth hormone secretion in monkeys. *J Clin Endocrinol Metab.* 1989;69(5):956–962.

244. Sortino MA, Cronin MJ, et al. Relaxin stimulates prolactin secretion from anterior pituitary cells. *Endocrinology.* 1989;124(4):2013–2015.

245. Granges G, Zilko P, et al. Fibromyalgia syndrome: assessment of the severity of the condition 2 years after diagnosis. *J Rheumatol.* 1994;21(3):523–552.

246. Bennett RM. Multidisciplinary group programs to treat fibromyalgia patients. *Rheum Dis Clin North Am.* 1996;22(2):351–367.

247. Bennett RM, Burckhardt CS, et al. Group treatment of fibromyalgia: a 6 month outpatient program. *J Rheumatol.* 1996;23(3):521–528.

248. Vlaeyen JW, Teeken-Gruben NJ, et al. Cognitive-educational treatment of fibromyalgia: a randomized clinical trial. I. Clinical effects. *J Rheumatol.* 1996; 23(7):1237–1245.

249. Goossens ME, Rutten-van Molken MP, et al. Cognitive-educational treatment of fibromyalgia: a randomized clinical trial. II. Economic evaluation. *J Rheumatol.* 1996;23(7):1246–1254.

250. Texidor MS. The nonpharmacological management of chronic pain via the interdisciplinary approach. In: Weiner RS, ed. *Pain Management: A Practical Guide for Clinicians.* Boca Raton, FL: St. Lucie Press; 1998:123–135.

251. Turk DC, Okifuji A, et al. Interdisciplinary treatment for fibromyalgia syndrome: clinical and statistical significance. *Arthritis Care Res.* 1998;11(3):186–195.

252. Turk DC, Okifuji A, et al. Differential responses by psychosocial subgroups of fibromyalgia syndrome patients to an interdisciplinary treatment. *Arthritis Care Res.* 1998;11(5):397–404.

253. Vasudevan SV, Lynch NT. Counseling the patient with chronic pain—the role of the physician. In: Lynch NT, Vasudevan SV, eds. *Persistent pain: Psychosocial Assessment and Intervention.* Boston: Kluwer Academic Publishers; 1998:117–132.

254. Rodway MR. Systems theory. In: Turner FJ, ed. *Social Work Treatment: Interlocking Theoretical Approaches.* New York: Free Press; 1986:514–539.

255. Bandura A, O'Leary A, et al. Perceived self-efficacy and pain control: opioid and nonopioid mechanisms. *J Pers Soc Psychol.* 1987;53(3):563–571.

256. Bandura A, Cioffi D, et al. Perceived self-efficacy in coping with cognitive stressors and opioid activation. *J Pers Soc Psychol.* 1988;55(3):479–488.

257. Bandura A. Exercise of personal and collective ef-

ficacy in changing societies. In: Bandura A, ed. *Self-Efficacy in Changing Societies.* Cambridge: Cambridge University Press; 1995:1–45.

258. Linton SJ. Chronic back pain: integrating psychological and physical therapy—an overview. *Behav Med.* 1994;20(3):101–104.

259. Linton SJ. Chronic back pain: activities training and physical therapy. *Behav Med.* 1994;20(3):105–111.

260. Keel PJ, Bodoky C, et al. Comparison of integrated group therapy and group relaxation training for fibromyalgia. *Clin J Pain.* 1998;14(3):232–238.

261. Bradley LA. Cognitive-behavioral therapy for primary fibromyalgia. *J Rheumatol Suppl.* 1989;19:131–136.

262. Nielson WR, Walker C, et al. Cognitive behavioral treatment of fibromyalgia syndrome: preliminary findings. *J Rheumatol.* 1992;19(1):98–103.

263. Goldenberg DL, Kaplan KH, et al. A controlled study of a stress-reduction, cognitive-behavioral treatment program in fibromyalgia. *J Musculoske Pain.* 1994; 2(2):53–66.

264. White KP, Nielson WR. Cognitive behavioral treatment of fibromyalgia syndrome: a followup assessment. *J Rheumatol.* 1995;22(4):717–721.

265. McCain GA. Treatment of the fibromyalgia syndrome. *J Musculoske Pain.* 1999;7(1/2):193–208.

266. Goldenberg DL. A review of the role of tricyclic medications in the treatment of fibromyalgia syndrome. *J Rheumatol Suppl.* 1989;19:137–139.

267. Carette S, McCain GA, et al. Evaluation of amitriptyline in primary fibrositis. A double-blind, placebo-controlled study. *Arthritis Rheum.* 1986;29(5):655–659.

268. Goldenberg DL, Felson DT, et al. A randomized, controlled trial of amitriptyline and naproxen in the treatment of patients with fibromyalgia. *Arthritis Rheum.* 1986;29(11):1371–1377.

269. Bennett RM, Gatter RA, et al. A comparison of cyclobenzaprine and placebo in the management of fibrositis. A double-blind controlled study. *Arthritis Rheum.* 1988;31(12):1535–1542.

270. Hamaty D, Valentine JL, et al. The plasma endorphin, prostaglandin and catecholamine profile of patients with fibrositis treated with cyclobenzaprine and placebo: a 5-month study. *J Rheumatol Suppl.* 1989;19: 164–168.

271. Scudds RA, McCain GA, et al. Improvements in pain responsiveness in patients with fibrositis after successful treatment with amitriptyline. *J Rheumatol Suppl.* 1989;19:98–103.

272. Quimby LG, Gratwick GM, et al. A randomized trial of cyclobenzaprine for the treatment of fibromyalgia. *J Rheumatol Suppl.* 1989;19:140–143.

273. Jaeschke R, Adachi J, et al. Clinical usefulness of amitriptyline in fibromyalgia: the results of 23 *N*-of-1 ran-

domized controlled trials. *J Rheumatol.* 1991;18(3): 447–451.

274. Reynolds WJ, Moldofsky H, et al. The effects of cyclobenzaprine on sleep physiology and symptoms in patients with fibromyalgia. *J Rheumatol.* 1991; 18(3):452–454.

275. Santandrea S, Montrone F, et al. A double-blind crossover study of two cyclobenzaprine regimens in primary fibromyalgia syndrome. *J Int Med Res.* 1993; 21(2):74–80.

276. Carette S, Bell MJ, et al. Comparison of amitriptyline, cyclobenzaprine, and placebo in the treatment of fibromyalgia. A randomized, double-blind clinical trial. *Arthritis Rheum.* 1994;37(1):32–40.

277. Russell IJ, Fletcher EM, et al. Treatment of primary fibrositis/fibromyalgia syndrome with ibuprofen and alprazolam. A double-blind, placebo-controlled study. *Arthritis Rheum.* 1991;34(5):552–560.

278. Hrycaj P, Stratz T, et al. Pathogenetic aspects of responsiveness to ondansetron (5-hydroxytryptamine type 3 receptor antagonist) in patients with primary fibromyalgia syndrome—a preliminary study. *J Rheumatol.* 1996;23(8):1418–1423.

279. Samborski W, Stratz T, et al. The 5-HT3 blockers in the treatment of the primary fibromyalgia syndrome: a 10-day open study with Tropisetron at a low dose. *Mater Med Pol.* 1996;28(1):17–19.

280. Moldofsky H, Lue FA, et al. The effect of zolpidem in patients with fibromyalgia: a dose ranging, double blind, placebo controlled, modified crossover study. *J Rheumatol.* 1996;23(3):529–533.

281. Biasi G, Manca S, et al. Tramadol in the fibromyalgia syndrome: a controlled clinical trial versus placebo. *Int J Clin Pharmacol Res.* 1998;18(1):13–19.

282. Bennett RM, Clark SC, et al. A randomized, double-blind, placebo-controlled study of growth hormone in the treatment of fibromyalgia. *Am J Med.* 1998;104(3):227–231.

283. American Physical Therapy Association. Guide to physical therapist practice. *Phys Ther.* 1997;77:1163–1650.

284. Burckhardt CS, Bjelle A. Education programmes for fibromyalgia patients: description and evaluation. *Baillières Clin Rheumatol.* 1994;8(4):935–955.

285. McCain GA, Bell DA, et al. A controlled study of the effects of a supervised cardiovascular fitness training program on the manifestations of primary fibromyalgia. *Arthritis Rheum.* 1988;31(9):1135–1141.

286. Mengshoel AM, Komnaes HB, et al. The effects of 20 weeks of physical fitness training in female patients with fibromyalgia. *Clin Exp Rheumatol.* 1992; 10(4):345–349.

287. Burckhardt CS, Mannerkorpi K, et al. A randomized,

controlled clinical trial of education and physical training for women with fibromyalgia. *J Rheumatol.* 1994;21(4):714–720.

288. Martin L, Nutting A, et al. An exercise program in the treatment of fibromyalgia. *J Rheumatol.* 1996;23(6): 1050–1053.

289. Wigers SH, Stiles TC, et al. Effects of aerobic exercise versus stress management treatment in fibromyalgia. A 4.5 year prospective study. *Scand J Rheumatol.* 1996;25(2):77–86.

290. Nichols DS, Glenn TM. Effects of aerobic exercise on pain perception, affect, and level of disability in individuals with fibromyalgia. *Phys Ther.* 1994;74(4): 327–332.

291. Nørregaard J, Lykkegaard JJ, et al. Exercise training in treatment of fibromyalgia. *J Musculoske Pain.* 1997; 5(1):71–79.

292. Dommerholt J. Posture. In: Tubiana R, Amadio P, eds. *Medical Problems of the Instrumentalist Musician.* London: Martin Dunitz; 2000:399–419.

293. Müller W, Kelemen J, et al. Spinal factors in the generation of fibromyalgia syndrome. *Z Rheumatol.* 1998; 57(Suppl 2):36–42.

294. Turk DC, Okifuji A, et al. Pain, disability, and physical functioning in subgroups of patients with fibromyalgia. *J Rheumatol.* 1996;23(7):1255–1262.

295. Deluze C, Bosia L, et al. Electroacupuncture in fibromyalgia: results of a controlled trial. *BMJ.* 1992; 305(6864):1249–1252.

296. Sandberg M, Lundeberg T, et al. Manual acupuncture in fibromyalgia: a long-term pilot study. *J Musculoske Pain.* 1999;7(3):39–58.

297. Gunn CC. 'Fibromyalgia'—"what have we created?" (Wolfe 1993) [letter]. *Pain.* 1995;60(3):349–350.

298. Dommerholt J. Fibromyalgia: time to consider a new taxonomy? *J Musculoske Pain.* 2000; 8(4):in press.

299. Bonica JJ. Definitions and taxonomy of pain. In: Bonica JJ, Loesser JD, et al, eds. *The Management of Pain.* Philadelphia: Lea & Febiger; 1990:18–27.

300. Russell IJ. Neurochemical pathogenesis of fibromyalgia syndrome. *J Musculoske Pain.* 1996;4(1/2):61–92.

301. Stanton-Hicks M, Jänig W, et al. Reflex sympathetic dystrophy: changing concepts and taxonomy. *Pain.* 1995;63(1):127–133.

302. Simons DG. Myofascial pain syndrome: one term but two concepts: a new understanding (editorial). *J Musculoskele Pain.* 1995;3(1):7–13.

303. Foreman PA. Temporomandibular joint and myofascial pain dysfunction—some current concepts. Part 1: Diagnosis. *N Z Dent J.* 1985;81(364):47–52.

304. Gunn CC. *The Gunn Approach to the Treatment of Chronic Pain.* 2nd ed. New York: Churchill Livingstone; 1997.

305. Simons DG, Travell JG, et al. *Myofascial Pain and Dysfunction: The Trigger Point Manual.* 2nd ed. Vol. 1, Baltimore: Williams & Wilkins; 1999.

306. Hong C-Z, Simons DG. Pathophysiologic and electrophysiologic mechanisms of myofascial trigger points. *Arch Phys Med Rehabil.* 1998;79(7):863–872.

307. Travell JG, Simons DG. *Myofascial Pain and Dysfunction: The Trigger Point Manual.* Vol. 2, Baltimore: Williams & Wilkins; 1992.

308. Russell IJ. Reliability of clinical assessment measures for the classification of myofascial pain syndrome. *J Musculoske Pain.* 1999;7(1/2):309–324.

309. Simons DG, Simons LS. Chronic Myofascial Pain Syndrome. In: Tollison CD, Satterthwaite JR, et al, eds. *Handbook of Pain Management.* Baltimore: Williams & Wilkins; 1994:556–577.

310. Rosomoff HL, Fishbain DA, et al. Physical findings in patients with chronic intractable benign pain of the neck and/or back. *Pain.* 1989;37(3):279–287.

311. Skootsky SA, Jaeger B, et al. Prevalence of myofascial pain in general internal medicine practice. *West J Med.* 1989;151:157–160.

312. Fricton JR. Myofascial pain syndrome: characteristics and epidemiology. *Adv Pain Res.* 1990;17:107–128.

313. Alfven G. The pressure pain threshold (PPT) of certain muscles in children suffering from recurrent abdominal pain of non-organic origin. An algometric study. *Acta Paediatr.* 1993;82(5):481–483.

314. Gerwin R. A study of 96 subjects examined both for fibromyalgia and myofascial pain (abstract). *J Musculoske Pain.* 1995;3(Suppl 1):121.

315. Gerwin RD. Differential diagnosis of myofascial pain syndrome and fibromyalgia. *J Musculoske Pain.* 1999; 7(1/2):209–215.

316. Baker BA. The muscle trigger: evidence of overload injury. *J Neurol Orthopaed Med Surg.* 1986;7(1): 35–44.

317. Dejung B. Die Verspannung des M. iliacus als Ursache lumbosacraler Schmerzen. *Manuelle Medizin.* 1987;25:73–81.

318. Tschopp K, Bachmann R. Das tempero-mandibuläre Myoarthropathiesyndrom—eine häufige Ursache für Gesichtsschmerzen. *Schweiz Rundsch Med Prax.* 1992; 81(15):468–472.

319. Dvorak J. Neurologische Ursachen für einen Handgelenkschmerz. *Orthopäde.* 1993;22(1):25–29.

320. Gerwin RD, Dommerholt J. Treatment of myofascial pain syndromes. In: Weiner R, ed. *Pain Management: A Practical Guide for Clinicians.* Boca Raton, FL: St. Lucie Press; 1997:217–229.

321. Gröbli C. Klinik und Pathophysiologie von myofaszialen Triggerpunkten. *Physiotherapie.* 1997;32(1): 17–26.

322. Imamura ST, Lin TY, et al. The importance of myofascial pain syndrome in reflex sympathetic dystrophy (or complex regional pain syndrome). In: Fischer AA, ed. *Myofascial Pain: Update in Diagnosis and Treatment.* Philadelphia: WB Saunders; 1997:207–211.

323. Novak CB, Mackinnon SE. Repetitive use and static postures: a source of nerve compression and pain. *J Hand Ther.* 1997;10:151–159.

324. Zohn DA. Relationship of joint dysfunction and soft-tissue problems. In: Fischer AA, ed. *Myofascial Pain: Update in Diagnosis and Treatment.* Philadelphia: WB Saunders; 1997:69–86.

325. Gerwin RD, Dommerholt, J. Myofascial trigger points in chronic cervical whiplash syndrome (abstract). *J Musculoske Pain.* 1998;6(Suppl 2):28.

326. Hendler NH, Kozikowski JG. Overlooked physical diagnoses in chronic pain patients involved in litigation. *Psychosomatics.* 1993;34(6):494–501.

327. Hendler NH, Zinreich J, et al. Three-dimensional CT validation of physical complaints in "psychogenic pain" patients. *Psychosomatics.* 1993;34:90–96.

328. Quintner JL, Cohen ML. Referred pain of peripheral nerve origin: an alternative to the "myofascial pain" construct. *Clin J Pain.* 1994;10(3):243–251.

329. Cannon WB, Rosenblueth A. *The Supersensitivity of Denervated Structures, a Law of Denervation,* New York: MacMillan; 1949.

330. Hong C-Z, Torigoe Y. Electrophysiological characteristics of localized twitch responses in responsive taut bands of rabbit skeletal muscle. *J Musculoske Pain.* 1994;2:17–43.

331. Hong C-Z. Pathophysiology of myofascial trigger point. *J Formos Med Assoc.* 1996;95(2):93–104.

332. Gerwin RD, Shannon S, et al. Interrater reliability in myofascial trigger point examination. *Pain.* 1997; 69(1–2):65–73.

333. Janda V. Muscle spasm: a proposed procedure for differential diagnosis. *J Manual Med.* 1991;6:136–139.

334. Mense S. Pathophysiologic basis of muscle pain syndromes. In: Fischer AA, ed. *Myofascial Pain: Update in Diagnosis and Treatment.* Philadelphia: WB Saunders; 1997:23–53.

335. Hong C-Z. Current research on myofascial trigger points—pathophysiological studies. *J Musculoske Pain.* 1999;7(1/2):121–129.

336. Wolfe F, Simons DG, et al. The fibromyalgia and myofascial pain syndromes: a preliminary study of tender points and trigger points in persons with fibromyalgia, myofascial pain syndrome and no disease. *J Rheumatol.* 1992;19(6):944–951.

337. Nice DA, Riddle DL, et al. Intertester reliability of judgments of the presence of trigger points in pa-

tients with low back pain. *Arch Phys Med Rehabil.* 1992;73(10):893–898.

338. Njoo KH, Van der Does E. The occurrence and inter-rater reliability of myofascial trigger points in the quadratus lumborum and gluteus medius: a prospective study in non-specific low back pain patients and controls in general practice. *Pain.* 1994;58(3):317–323.

339. Lew PC, Lewis J, et al. Inter-therapist reliability in locating latent myofascial trigger points using palpation. *Manual Therapy.* 1997;2(2):87–90.

340. Hong C-Z, ChenY-N, et al. Pressure threshold for referred pain by compression on the trigger point and adjacent areas. *J Musculoske Pain.* 1996;4(3):61–79.

341. Fischer AA. Pressure threshold measurement for diagnosis of myofascial pain and evaluation of treatment results. *Clin J Pain.* 1986;2(4):207–214.

342. Bendtsen L, Jensen R, et al. Muscle palpation with controlled finger pressure: new equipment for the study of tender myofascial tissues. *Pain.* 1994;59(2):235–239.

343. Fenger-Grøn LS, Graven-NielsenT, et al. Muscular sensibility assessed by electrical stimulation and mechanical pressure. *J Musculoske Pain.* 1998;6(4):33–44.

344. Cummings GS, Tillman LJ. Remodeling of dense connective tissue in normal adult tissues. In: Currier DP, Nelson RM, eds. *Dynamics of Human Biologic Tissues.* Philadelphia: FA Davis; 1992:45–73.

345. Mense S, Skeppar RF. Discharge behavior of feline gamma-motoneurons following induction of an artificial myositis. *Pain.* 1991;46:201–210.

346. Gerwin RD. Neurobiology of the myofascial trigger point. *Baillières Clin Rheumatol.* 1994;8(4):747–762.

347. Wilson PR, Lamer TJ. Pain mechanisms: anatomy and physiology. In: Raj PP, ed. *Practical Management of Pain.* St. Louis: Mosby Year Book; 1992:65–80.

348. Dommerholt J, Gröbli C. Knee pain. In: Whyte-Ferguson L, Gerwin RD, eds. *Clinical Mastery of Myofascial Pain Syndrome.* Baltimore: Lippincott, Williams & Wilkins; in press.

349. Fowler PJ, Lubliner J. Functional anatomy and biomechanics of the knee joint. In: Grifin LY, ed. *Rehabilitation of the Injured Knee.* St. Louis: Mosby; 1995:7–19.

350. Fischer AA. Treatment of myofascial pain. *J Musculoske Pain.* 1999;7(1/2):131–142.

351. Gerwin RD, Gevirtz R. Chronic myofascial pain: iron insufficieny and coldness as risk factors. *J Musculoske Pain.* 1995;3(Suppl 1):120.

352. Cassisi JE, Sypert GW, et al. Pain, disability, and psychological functioning in chronic low back pain subgroups: myofascial versus herniated disc syndrome. *Neurosurgery.* 1993;33(3):379–385.

353. Faucett JA. Depression in painful chronic disorders: the role of pain and conflict about pain. *J Pain Symptom Manage.* 1994;9(8):520–526.

354. Zautra AJ, Marbach JJ, et al. The examination of myofascial face pain and its relationship to psychological distress among women. *Health Psychology.* 1995;14: 223–231.

355. Hey LR, Helewa A. Myofascial pain syndrome: a critical review of the literature. *Physiother Can.* 1994; 46(1):28–36.

356. Bohr T. Problems with myofascial pain syndrome and fibromyalgia syndrome [editorial]. *Neurology.* 1996; 46(3):593–597.

357. Cohen ML. Arthralgia and myalgia. In: Campbell JN, Ed. *Pain 1996—An Updated Review.* Seattle: IASP Press; 1996:327–337.

358. Weeks VD, Travell J. How to give painless injections. *AMA Scientific Exhibits.* New York: Grune & Stratton; 1957:318–322.

359. Hubbard, DR, Berkoff GM. Myofascial trigger points show spontaneous needle EMG activity. *Spine.* 1993; 18:1803–1807.

360. Simons DG, Hong C-Z, et al. Prevalence of spontaneous electrical activity at trigger spots and control sites in rabbit muscle. *J Musculoske Pain.* 1995;3:35–48.

361. Simons DG, Hong C-Z, et al. Nature of myofascial trigger points, active loci (abstract). *J Musculoske Pain.* 1995;3(Suppl 1):62.

362. Hendler N, Fink H, et al. Myofascial syndrome: response to trigger-point injections. *Psychosomatics.* 1983;24:990–999.

363. Donaldson CCS, Skubick DL, et al. The evaluation of trigger-point activity using dynamic EMG techniques. *Am J Pain Management.* 1994;4:118–122.

364. Simons DG, Dexter JR. Comparison of local twitch responses elicited by palpation and needling of myofascial trigger points. *J Musculoske Pain.* 1995;3:49–61.

365. Fischer AA. New approaches in treatment of myofascial pain. In: Fischer AA, ed. *Myofascial Pain: Update in Diagnosis and Treatment.* Philadelphia: WB Saunders; 1997:153–170.

366. Gerwin RD, Duranleau D. Ultrasound identification of the myofacial trigger point. *Muscle Nerve.* 1997;20(6): 767–768.

367. Lewis J, Tehan P. A blinded pilot study investigating the use of diagnostic ultrasound for detecting active myofascial trigger points. *Pain.* 1999;79(1):39–44.

368. Mense S. Nociception from skeletal muscle in relation to clinical muscle pain. *Pain.* 1993;54:241–289.

369. Vecchiet L, Dragani L, et al. Experimental referred pain and hyperalgesia from muscles in humans. In: Vecchiet L, Albe-Fessard D, et al, eds. *New Trends in Referred Pain and Hyperalgesia.* Amsterdam: Elsevier Science Publishers; 1993:239–249.

370. Vecchiet L, Giamberardino,MA. Referred pain: clinical significance, pathophysiology and treatment. In: Fischer AA, ed. *Myofascial Pain: Update in Diagnosis and Treatment*. Philadelphia: WB Saunders; 1997: 119–136.

371. Mense S, Meyer H. Different types of slowly conducting afferent units in cat skeletal muscle and tendon. *J Physiol*. 1985(363):403–417.

372. Torebjörk HE, Ochoa JL, et al. Referred pain from intraneural stimulation of muscle fascicles in the median nerve. *Pain*. 1984;18:145–156.

373. Dwyer A, Aprill C, et al. Cervical zygopophyseal joint pain patterns, 1: a study in normal volunteers. *Spine*. 1990;15(6):453–457.

374. Neumann M. Trunk pain. In: Raj PP, ed. *Practical Management of Pain*. St. Louis: Mosby Year Book; 1992:258–271.

375. Scudds RA, Landry M, et al. The frequency of referred signs from muscle pressure in normal healthy subjects (abstract). *J Musculoske Pain*. 1995;3(Suppl 1):99.

376. Bellew JW. Lumbar facets: an anatomic framework for low back pain. *J Manual Manipulative Therapy*. 1996;4(4):149–156.

377. Gunn CC. Radiculopathic pain: diagnosis, treatment of segmental irritation or sensitization. *J Musculoske Pain*. 1997;5(4):119–134.

378. Janda V. Muscles and motor control in cervicogenic disorders; assessment and management. In: Grant R, ed. *Physical Therapy of the Cervical and Thoracic Spine*. New York: Churchill Livingstone; 1994:195–216.

379. Abdel-Fattah RA. *Preventing Temporomandibular Joint (TMJ) and Odontostomatognatic (OSGS) Injuries in Dental Practice*. Boca Raton, FL: CRC Press; 1993.

380. Jaeger B. Overview of head and neck region. In: Simons DG, Travell JG, et al, eds. *Myofascial Pain and Dysfunction: The Trigger Point Manual*. Baltimore: Williams & Wilkins; 1999:237–277.

381. Fricton JR, Kroening R, et al. Myofascial pain syndrome of the head and neck: a review of clinical characteristics of 164 patients. *Oral Surg Oral Med Oral Pathol*. 1985;60(6):615–623.

382. Andersen JH, Kærgaard A, et al. Myofascial pain in different occupational groups with monotonous repetitive work (abstract). *J Musculoske Pain*. 1995;3(Suppl 1):57.

383. Lin TY, Teixeira MJ, et al. Work-related musculoskeletal disorders. In: Fischer AA, Ed. *Myofascial Pain: Update in Diagnosis and Treatment*. Philadephia: WB Saunders; 1997:113–118.

384. Farrell J, Littlejohn G. Association between task performance and tender point pain threshold to pressure in normal subjects. *J Musculoske Pain*. 1997; 5(1):19–47.

385. Feuerstein M, Hickey PF. Ergonomic approaches in the clinical assessment of occupational musculoskeletal disorders. In: Turk DC, Melzack R, eds. *Handbook of Pain Assessment*. New York: The Guilford Press; 1992:71–99.

386. Kuorinka I, Forcier L. *Work Related Musculoskeletal Disorders (WMSDs): A Reference Book for Prevention*. Bristol: Taylor & Francis, Inc; 1995.

387. Skubick DL, Clasby R, et al. Carpal tunnel syndrome as an expression of muscular dysfunction in the neck. *J Occupational Rehab*. 1993;3(1):31–43.

388. Hünting W, Läubli T, et al. Postural and visual loads at VDT workplace: 1. constrained postures. *Ergonomics*. 1981;24(12):917–931.

389. Silverstein BA, *The Prevalence of Upper Extremity Cumulative Trauma Disorders in Industry*. Ann Arbor: University of Michigan; 1985.

390. Amano M, Umeda G, et al. Characteristics of work actions of shoe manufacturing assembly line workers and a cross sectional factor control study on occupational cervicobrachial disorders. *Jpn J Ind Health*. 1988;30(1):3–12.

391. Rosen NB. Myofascial pain: the great mimicker and potentiator of other diseases in the performing artist. *Md Med J*. 1993;42(3):261–266.

392. Dommerholt J, Norris RN, et al. Therapeutic management of the instrumental musician. In: Sataloff RT, Brandfonbrener AG, et al, eds. *Performing Arts Medicine*. San Diego: Singular Publishing Group; 1998: 277–290.

393. Glogowsky C, Wallraff J. Ein Beitrag zur Klinik und Histologie der Muskelhärten (Myogelosen). *Z Orthop*. 1951;80:237–268.

394. Fassbender HG. Morphologie und pathogenese des weichteilrheumatismus. *Z Rheumaforsch*. 1973;32: 355–374.

395. Fassbender HG. *Psyche und Rheuma: psychosomatische Schmerzsyndrome des Bewegungsapparates*, Basel: Schwabe/Eular Publ; 1975:75–86.

396. Gariphianova MB. The ultrastructure of myogenic trigger points in patients with contracture of mimetic muscles (abstract). *J Musculoske Pain*. 1995;3(Suppl 1):23.

397. Reitinger A, Radner H, et al. Morphologische Untersuchung an Triggerpunkten. *Manuelle Medizin*. 1996; 34:256–262.

398. Pongratz DE, Späth M. Morphologic aspects of muscle pain syndromes. In: Fischer AA, ed. *Myofascial Pain: Update in Diagnosis and Treatment*. Philadelphia: WB Saunders; 1997:55–68.

399. Brückle W, Sückfull M, et al. Gewebe-pO2-Messung

in der verspannten Rückenmuskulatur (m. erector spinae). *Z. Rheumatol.* 1990;49:208–216.

400. Martonosi AN. Regulation of calcium by the sarcoplasmic reticulum. In: Engel AG, Franzini-Armstrong C, eds. *Myology.* New York: McGraw-Hill; 1994:553–584.

401. Shenoi R, Nagler W. Trigger points related to calcium channel blockers (letter). *Muscle Nerve.* 1996;19(2):256.

402. Ito Y, Miledi R, et al. Transmitter release induced by a 'factor' in rabbit serum. *Proc R Soc Lond B Biol Sci.* 1974;187:235–241.

403. Simons D. Clinical and etiological update of myofascial pain from trigger points. *J Musculoske Pain.* 1996;4(1/2):93–121.

404. Hong C-Z, Yu J. Spontaneous electrical activity of rabbit trigger spot after transection of spinal cord and peripheral nerve. *J Musculoske Pain.* 1998;6(4):45–58.

405. Porter CW, Barnard EA. Ultrastructural studies on the acetylcholine receptor at motor end plates of normal and pathologic muscles. *Ann N Y Acad Sci.* 1976;274:85–107.

406. Grohovaz F, Lorenzon P, et al. Properties of acetylcholine receptors in adult rat skeletal muscle fibers in culture. *J Membr Biol.* 1993;136(1):31–42.

407. Hong C-Z. Pathophysiology of myofascial trigger point. *J Formos Med Assoc.* 1996;95(2):93–104.

408. Hong C-Z. Myofascial trigger point injection. *Crit Rev Phys Med Rehabil.* 1993;5(2):203–217.

409. Acquadro MA, Borodic GE. Treatment of myofascial pain with botulinum A toxin [letter]. *Anesthesiology.* 1994;80(3):705–706.

410. Cheshire WP, Abashian SW, et al. Botulinum toxin in the treatment of myofascial pain syndrome. *Pain.* 1994;59(1):65–69.

411. Yue SK. Initial experience in the use of botulinum toxin A for the treatment of myofascial related muscle dysfunctions. *J Musculoske Pain.* 1995;3(Suppl.1):22.

412. Alo KM, Yland MJ, et al. Botulinum toxin in the treatment of myofascial pain. *Pain Clinic.* 1997;10(2):107–116.

413. Raj PP. Botulinum toxin in the treatment of pain associated with musculoskeletal hyperactivity. *Curr Rev Pain.* 1997;1:403–416.

414. Wheeler AH, Goolkasian P, et al. A randomized, double-blind, prospective pilot study of botulinum toxin injection for refractory, unilateral, cervicothoracic, paraspinal, myofascial pain syndrome. *Spine.* 1998;23(15):1662–1666.

415. de Paiva A, Meunier FA, et al. Functional repair of motor endplates after botulinum neurotoxin type A poisoning: biphasic switch of synaptic activity between nerve sprouts and their parent terminals. *Proc Natl Acad Sci U S A.* 1999;96(6):3200–3205.

416. Lewis C, Gevirtz R, et al. Needle trigger point and surface frontal EMG measurements of psychophysiological responses in tension-type headache patients. *Biofeedback & Self-Regulation.* 1994;3:274–275.

417. McNulty WH, Gevirtz RN, et al. Needle electromyographic evaluation of trigger point response to a psychological stressor. *Psychophysiology.* 1994;31(3):313–316.

418. Hubbard DR. Chronic and recurrent muscle pain: pathophysiology and treatment, and review of pharmacologic studies. *J Musculoske Pain.* 1996;4:123–143.

419. Chen JT, Chen SM, et al. Phentolamine effect on the spontaneous electrical activity of active loci in a myofascial trigger spot of rabbit skeletal muscle. *Arch Phys Med Rehabil.* 1998;79(7):790–794.

420. Simons DG, Hong C-Z, et al. Spike activity in trigger points. *J Musculoske Pain.* 1995;3(Suppl 1):125.

421. Partanen J. End plate spikes in the human electromyogram. Revision of the fusimotor theory. *J Physiol Paris.* 1999;93(1–2):155–166.

422. Grassi C, Passatore M. Action of the sympathetic system on skeletal muscle. *Ital J Neurol Sci.* 1988;9(1):23–28.

423. Barker D, Saito M. Autonomic innervation of receptors and muscle fibres in cat skeletal muscle. *Proc R Soc Lond B Biol Sci.* 1981;212(1188):317–332.

424. Ljung BO, Forsgren S, et al. Sympathetic and sensory innervations are heterogeneously distributed in relation to the blood vessels at the extensor carpi radialis brevis muscle origin of man. *Cell Tissue Org.* 1999;165(1):45–54.

425. Jänig W. The sympathetic nervous system in pain: physiology and pathophysiology. In: Stanton-Hicks M, ed. *Pain and the Sympathetic Nervous System.* Boston: Kluwer Academic Publishers; 1990:17–89.

426. Vecchiet L, Giamberardino MA, et al. Latent myofascial trigger points: changes in muscular and subcutaneous pain thresholds at trigger point and target level. *J Manual Medicine.* 1990;5:151–154.

427. Vecchiet L, Giamberardino MA, et al. Comparative sensory evaluation of parietal tissues in painful and nonpainful areas in fibromyalgia and myofascial pain syndrome. In: Gebhart GF, Hammond DL, et al, eds. *Proceedings of the 7th World Congress on Pain (Progress in Pain Research and Management).* Seattle: IASP Press; 1994:177–185.

428. Mense S. Referral of muscle pain: new aspects. *Amer Pain Soc J.* 1994;3:1–9.

429. Marchettini P, Simone DA, et al. Pain from excitation of identified muscle nociceptors in humans. *Brain Res.* 1996;740(1–2):109–116.

430. Bendtsen L, Jensen R, et al. Qualitatively altered nociception in chronic myofascial pain. *Pain.* 1996;65: 259–264.

431. Hoheisel U, Mense S, et al. Appearance of new receptive fields in rat dorsal horn neurons following noxious stimulation of skeletal muscle: a model for referral of muscle pain? *Neurosci Lett.* 1993;153:9–12.

432. Hoheisel U, Sander B, et al. Myositis-induced functional reorganisation of the rat dorsal horn: effects of spinal superfusion with antagonists to neurokinin and glutamate receptors. *Pain.* 1997;69:219–230.

433. Siddall PJ, Cousins MJ. Spine update; spinal pain mechanisms. *Spine.* 1997;22(1):98–104.

434. Ruch TC. Pathophysiology of pain. In: Ruch TC, Patton HD, eds. *Physiology and Biophysics: The Brain and Neural Function.* Philadelphia: WB Saunders; 1979: 272–324.

435. Hu JW, Sessle BJ, et al. Stimulation of craniofacial muscle afferents induces prolonged facilitatory effects in trigeminal nociceptive brainstem neurons. *Pain.* 1992;48:53–60.

436. McMahon SB, Wall PD. Physiological evidence for branching of peripheral unmyelinated sensory afferent fibers in the rat. *J Comp Neurol.* 1987;261:130–136.

437. McMahon SB, Wall PD. Functional significance of multiple branches of afferent C fibres in peripheral nerve. In: Schmidt RF, Schaible H-G, et al, eds. *Fine Afferent Nerve Fibers and Pain.* Weinheim: VCH; 1987: 97–104.

438. Saal JA, Saal JS. Rehabilitation of the patient. In: White AH, Anderson R, eds. *Conservative Care of Low Back Pain.* Baltimore: Williams & Wilkins; 1991: 21–34.

439. Hopwood MB, Abram, SE. Factors associated with failure of trigger point injections. *Clin J Pain.* 1994; 10(3):227–234.

440. Nelson LS, Hoffman RS. Intrathecal injection: unusual complication of trigger-point injection therapy. *Ann Emerg Med.* 1998;32(4):506–508.

441. Carlson CR, Okeson JP, et al. Reduction of pain and EMG activity in the masseter region by trapezius trigger point injection. *Pain.* 1993;55(3):397–400.

442. Gröbli C, Dommerholt J. Myofasziale Triggerpunkte; Pathologie und Behandlungsmöglichkeiten. *Manuelle Medizin.* 1997;35:295–303.

443. Banks SL, Jacobs DW, et al. Effects of authogenic relaxation training on electromyographic activity in active myofascial trigger points. *J Musculoske Pain.* 1998;6(4):23–32.

444. Dvorák J, Dvorák V. *Manual Medicine; diagnostics,* Stuttgart: Georg Thieme Verlag; 1990.

445. Butler DS, Shacklock MO, et al. Treatment of altered nervous system mechanics. In: Boyling JD, Palastanga N, eds. *Grieve's Modern Manual Therapy.* Edinburgh: Churchill Livingstone; 1994:693–703.

446. Sahrmann SA. Adult posturing. In: Kraus S, ed. *TMJ Disorders: Management of the Craniomandibular Complex.* New York: Churchill Livingstone; 1988: 295–309.

447. Carriere B. Therapeutic exercise and self-correction programs. In: Flynn TW, ed. *The Thoracic Spine and Rib Cage: Musculoskeletal Evaluation and Treatment.* Boston: Butterworth-Heinemann; 1996:287–307.

PART III

Evaluation and Treatment of the Myofascial System

CHAPTER 7

Basic Evaluation of the Myofascial System

Robert I. Cantu and Alan J. Grodin

This chapter offers the clinician information and insight into the evaluation of the myofascial system. Although other aspects of the biomechanical evaluation of the spine may be discussed when appropriate, the main focus remains on the myofascial system. Myofascial assessment represents only one aspect of the total evaluation, and the results should always be correlated with other findings to assess accurately the functional (or dysfunctional) status of the spine and/or extremities.

Dysfunction is defined by Dorland's as "a disturbance, impairment, or abnormality of the functioning."[1] More specifically, somatic dysfunction can be defined as "impaired or altered function of related components of the somatic system. Somatic dysfunction is a state of altered mechanics, palpable changes of integrity, increased or decreased mobility and autonomic changes."[2] A therapist diagnoses dysfunction in the same manner that a physician diagnoses pathology: correlation of findings. When a physician is looking for pathology in relation to low-back pain, the diagnosis is not made based on radiology or physical examination alone. In the case of discogenic pathology, for example, the physician uses the history, physical examination, radiologic findings, and electromyograms (EMGs) in order to determine if true discogenic radiculopathy exists. If the patient has an m resonance imaging (MRI) with a positive finding

for discogenic lesion, in the absence of any other finding, the herniation may not be the cause of the pain and dysfunction. The physician who would diagnose discogenic pathology on the basis of MRI alone would be premature in making the diagnosis. If, however, the patient is experiencing low-back pain, has referred pain in the lower extremity, has diminished reflexes, selective muscle weakness, and positive EMG and MRI results, the findings together definitively correlate for discogenic pathology.

The physical therapist also diagnoses significant dysfunction in the same way. All findings from the history, visual, palpatory, and movement examinations are correlated to determine dysfunction. Postural asymmetry caused by a leg-length discrepancy in itself is not dysfunctional. Active movement abnormalities alone are not necessarily dysfunctional. Segmental hypermobility and hypomobility in and of themselves are not necessarily dysfunctional. Connective tissue changes in the absence of other findings are not dysfunctional. If several findings from the evaluation are abnormal, however, a strong statement can be made for dysfunction. For example, a patient may have symptoms including localized unilateral low-back pain, a postural fulcrum at L4–5, an exaggerated lumbar curve reversal on forward bending (with a fulcrum of motion at L4–5), tenderness to palpation at the L4–5 interspace, increased erector spinae muscle tone in the lumbar spine, hypermobility of the L4–5 segment, and increased connective tissue in the area. In this theoretical scenario (rarely this clean-cut), a dysfunction of L4–5 exists, with L4–5 hypermobility, movement imbalance as a result of the hypermobility, increased connective tissue in the area as the body's attempt to stabilize it, and protective muscle guarding with altered muscular recruitment patterns. Although none of the above abnormalities alone would have constituted dysfunction, the combination of abnormalities does. Treatment can be initiated by addressing this combination of factors that contribute to the overall dysfunction.

The aspects of myofascial evaluation considered in this chapter are the history, postural and structural evaluation, movement analysis, palpatory examination, and passive motion analysis. The myofascial aspects of this evaluation are stressed.

HISTORY

Cyriax stated that the history is of great importance, especially in spinal conditions.[3] Most clinicians have a standardized routine questionnaire and historical format, but several key questions should always be asked when looking for myofascial-type pain syndromes.

1. What is the quality of the pain? Myofascial pain is usually dull and aching, as well as poorly localized. If the patient is reporting specific, sharp pain, which is easily reproduced, specific pathology may exist rather than a myofascial-type syndrome.

2. How is the patient sleeping at night? Again, one of the critical factors in myofascial pain is the disturbed sleep pattern. Typically, the patient will report difficulty going to sleep and frequent awakenings during the night. Patients usually report feeling unrefreshed and fatigued in the morning.

3. Is the pain waking the patient? The loss of sleep is not due to sharp pain awakening the patient, but specifically, to poor, interrupted, and nonrefreshing sleep. If the patient is awakening, for example, because sharp pain occurs with movement, the sleep interruption is not as significant and is usually indicative of specific pathology, not myofascial pain.

4. How much generalized fatigue is the patient experiencing during the day? Ninety percent to 100% of patients with fibromyalgia report feelings of fatigue during the day, even if they are relatively inactive.[4]

5. What pattern does the pain follow during the day? A typical daytime pattern for myofascial pain is increased stiffness and pain in the early morning, with a slight drop-off in symptoms at midmorning, and with the pain remaining somewhat constant throughout the day. Increased activity will usually aggravate the condition, but the symptoms remain global and diffuse.

6. What medications is the patient taking? This is extremely important if a myofascial-type pain syndrome is suspected, since few drugs have proven to be even slightly helpful; some drugs can actually be detrimental. Valium, for example, can help the patient sleep, but the drug can actually block stage 4 sleep. The drugs of choice for restoring normal sleep patterns are amitriptyline (Elavil[R]) and cyclobenzaprine (Flexeril[R]).

7. Does the patient have a hyperallergenic history or have a tendency toward irritable bowel syndrome? Many patients with fibromyalgia also have problems with allergies, histamine reactions, or irritable bowels. All remaining questions for the patient history should be asked as a matter of routine, but the above questions should be emphasized when a myofascial-type pain syndrome is suspected.

POSTURAL AND STRUCTURAL EVALUATION

The first part of any objective evaluation for somatic dysfunction consists of observing posture. Posture can be defined as balance and muscular coordination and adaptation with minimal expenditure of energy. It is the position the body assumes in preparation for the next movement; it is not necessarily a static position.[5] Posture is dynamic, requiring muscular forces and creation of connective tissue tensions. Looking at the skeletal aspects of posture without considering the dynamic aspects gives a shallow, incomplete picture of the postural influences of dysfunction. Body posture may give preliminary clues to the location of a movement disturbance or to an area where stress may occur due to overuse or trauma. Posture observation directs the clinician's focus

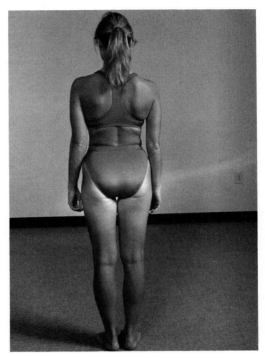

Figure 7–2

on a particular area or areas of the system that may be significantly dysfunctional.

Observation of Posture

The patient should be viewed from posterior, anterior, and lateral angles to ensure accurate assessment (Figures 7–1 through 7–3). In integrating the myofascial system into postural evaluation, the clinician should look for muscle asymmetry, connective tissue asymmetry, and increased muscular activity that may correlate with abnormal structural deviations. The entire body should be viewed, from the subcranial area down to the feet, since the fascial planes can be restricted over large areas of the body.

Muscle asymmetry may be a result of prolonged shortening or lengthening of a muscle group, due, for example, to a leg-length discrepancy or a pelvic obliquity. Connective tissue

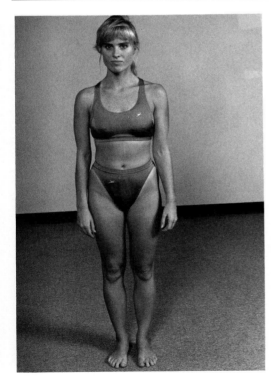

Figure 7–1

Figure 7–3

of myofascial disequilibrium in the spine is the dysfunction caused by the forward-head posture.

In the forward-head posture, the midcervical facet joints are in the "up and forward position" or forward bent. There is generally a loss of lordosis in this area, with a tendency toward hypermobility (Figures 7–4, 7–5, and 7–6). In the upper cervical and subcranial area, the facet joints are in the "down and back" position or backward bent in order to compensate for the forward bending in the lower cervical spine and to keep the eyes in horizontal. This creates compression of the facet joints, which can lead to hypomobility and a shortening of the posterior myofascial structures. Because the greater occipital nerve pierces the subcranial myofascia, compression of this nerve can create occipital and frontal headaches. The anterior cervical spine compensates by lengthening, changing the length-tension relationships, and contributing to a weakness in the area.

asymmetry may be due to abnormal stresses applied to an area, creating a localized proliferation of connective tissue, as in a spondylolisthesis. Increased muscular activity is usually a precursor to muscle asymmetry and is usually found in more acute cases.

While observing body asymmetry is important, the clinician must remember that the human body is, by nature's design, asymmetrical. Hand, leg, and eye dominance possibly contributes to myofascial and structural asymmetry. The critical factor in determining whether or not the asymmetry is significant is its correlation to other relevant evaluative findings.

Postural observations give the clinician some insights into the overall equilibrium of the spine. When looking at joint equilibrium in the spine, consider that a joint can be stable and in optimal functional position only if there is equilibrium between the forces acting on it. A good example

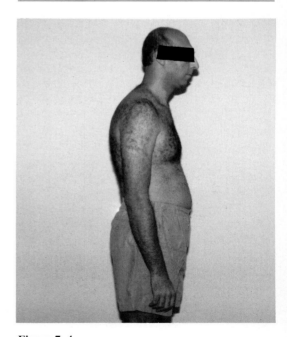

Figure 7–4

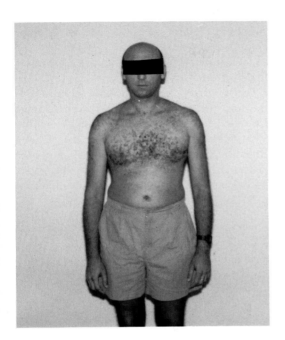

Figure 7–5

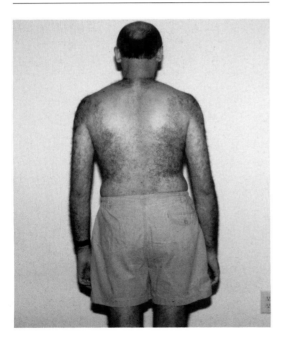

Figure 7–6

In the forward-head posture, the mandible tends to open, so the masseters and temporalis are engaged to keep the mouth closed. This leads to new, but abnormal, hyperactive muscle patterns, where the muscles become facilitated and can create dysfunctions such as nocturnal bruxism. This can lead to eventual degenerative changes in the temporomandibular joint.

In the upper thoracic area, the facets are again in a forward bent position, with the posterior myofascial structures on a stretch. In the anterior chest wall, the myofascial structures are held in a shortened position. The shoulder girdle complex is held in a protracted position with the glenohumeral joint tending to go toward internal rotation. Because the anterior thorax is held in a shortened position, diaphragmatic breathing is compromised and the accessory muscles of respiration are facilitated, leading to a potentially elevated first rib, a compromise of the costoclavicular space, and increasing susceptibility to thoracic outlet-type symptomatology. The lumbar spine can be either hyperlordotic or hypolordotic. If hypolordotic, a stretching of posterior structures occurs, resulting in hypermobility and possible strain on the posterior aspect of the disc (Table 7–1).

Myofascial Aspects

The myofascial aspects of the forward head posture correlate well with the mechanical aspects. The work of Janda[2,6] has helped tremendously in correlating the effects of myofascial imbalances on postural imbalances. The principles he put forth include the relationship of "postural" and "phasic" muscles and their correlation to agonist/antagonist muscle groups. In histological terms, postural and phasic muscles are differentiated by oxidative capacity and ability to generate large or small amounts of force for short or long periods of time. The terms "postural" and "phasic," in the context of Janda's work and for the purposes of this discussion, relate more to how the muscle responds to dysfunction. In the myofascial context, a postural muscle is one that responds to dysfunction or abnormal stress by tightening, whereas a phasic

Table 7–1 Postural Sequence for the Forward-Head Posture

Forward bending of the midcervical facet joints

Backward bending (extension) of the occiput atlas

Shortening of suboccipital muscles, resulting in potential impingement of the greater or lesser occipital nerves

Imbalance between the sternocleidomastoid, the levator scapula, and the trapezius

Imbalance between the anterior cervical musculature (including the suprahyoid and infrahyoid muscles) and posterior cervical extensors

Shoulder girdle protraction with internal rotation (the latissimus, subscapularis, pectoralis, and teres major being involved)

Increased thoracic kyphosis with decreased lumbar lordosis

Increased activity of the accessory respiratory muscles due to poor diaphragmatic breathing and poor expansion of the lower rib cage

Elevation of the first rib by increased scalene activity

Anterior and posterior restriction of the first rib articulations

Tendency toward thoracic outlet symptomatology

Cervical imbalance with a tendency toward degenerative joint disease from C5 through C7

Muscular imbalance leading to abnormal muscle firing (some muscles become facilitated with trigger points)

Joints and soft tissues maintained in shortened range lead to restriction of joint capsules and loss of proprioception

muscle is one that responds to dysfunction by weakening. In the agonist/antagonist scheme, usually one muscle or set of muscles responds to dysfunction by weakening while the other responds by tightening. An obvious example of this is the quadriceps and hamstrings. The quadriceps rarely become tight, whereas the hamstrings tend to tighten on a regular basis. If the knee is injured, the quadriceps usually weaken and atrophy, while the hamstrings rarely show significant atrophy or weakness. These agonist/antagonist relationships play a vital role in postural problems of the spine.

The forward-head posture once again can be used as a clinical example, being by far the most common presentation in the clinic. A smaller percentage of patients do, however, have axially extended posture. When one superimposes the myofascial elements onto the arthrokinematics of dysfunctional posture, strong correlations can be made (Figures 7–7, 7–8, and 7–9).

Cervical Spine

In the forward-head posture, the cervical lordosis is increased and the straight-line distance between the occiput and the cervicothoracic junction is decreased. This relationship places the cervical erector spinae in a shortened position, which over a period of time permanently shortens the muscle. This is especially true in the upper cervical spine. In the myofascial scheme, the cervical erector spinae are classified as postural muscles, which respond to dysfunction by tightening, facilitating the dysfunction. The anterior musculature, on the other hand, is in an elongated position, which over a period of time

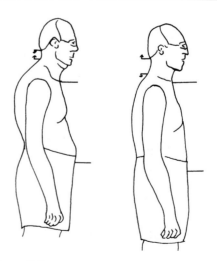

Figure 7–7

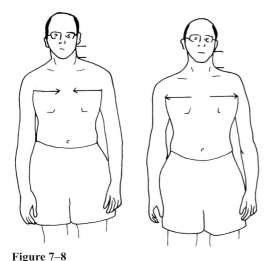

Figure 7–8

Table 7–2 Cervical/Upper Thoracic Agonist/ Antagonist Relationships

Postural	Phasic
Upper trapezius	Latissimus dorsi
Levator scapulae	
Pectoralis major	Mid/lower trapezius
(upper part)	
Pectoralis minor	Rhomboids
Cervical erector	Anterior cervical
spinae	musculature

creates a permanent lengthening. Because the muscle group responds to dysfunction by weakening, the forward-head posture is further enhanced (Table 7–2).

Thoracic Spine

In the forward-head posture, there is an increased kyphosis of the thoracic spine. The

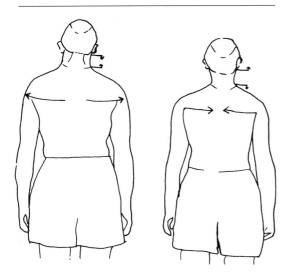

Figure 7–9

straight-line distance between the manubrium and the umbilicus, as well as the straight-line distance between glenohumeral joints, is decreased. This places the pectoralis major and minor, along with the upper trapezius, in a shortened position. In the myofascial system, the pectoralis major and minor muscles respond to dysfunction by tightening, as does the upper trapezius. The middle and lower trapezius and rhomboid muscles weaken in response to dysfunction, which further facilitates the thoracic dysfunction. Once again, antagonistic muscle groups respond in opposite ways to facilitate the same dysfunction. As noted, the anterior of the diaphragm, which, in turn, facilitates the upper thoracic accessory breathing muscles, further compound the problem (Table 7–3).

Lumbar Spine

In the lumbar spine, two situations commonly exist. The first, excessive lumbar lordosis, can be correlated to dysfunctional muscle groups. The increased lumbar lordosis includes a tightening of the lumbar erector spinae, psoas muscle groups, iliacus, and tensor fasciae latae. The antagonistic groups, which include the abdominals and the gluteus maximus, weaken, further facilitating the dysfunction. Corresponding joint dysfunction includes hypomobility of the lumbar segments, with tightening of the posterior structures (Figure 7–10, Tables 7–4 and 7–5).

The other scenario, in which there is a loss of lumbar lordosis, pits the hamstrings and pos-

Table 7–3 Muscle Agonist/Antagonist Groups of the Cervicothoracic Area with Resulting Dysfunctions

Muscle Group	Action	Response to Dysfunction	Results of Dysfunction
Upper trapezius levator scapulae	—elevation of shoulder girdle —assist in adduction of scapula —BB and SB of spine	Tightens	—elevation/adduction of scapula —increased cervical lordosis —restricted axial extension —limited side bending and rotation of cervical spine
Pectoralis major (upper part)	—shoulder flexion —horizontal adduction of humerus	Tightens	—restricted shoulder flexion —restricted horizontal adduction
Pectoralis minor	—protraction of scapula —accessory breathing muscle	Tightens	—scapular abduction with outward rotation of inferior angle —winging of inferior border of scapula —increased thoracic kyphosis
Rhomboids middle/lower trapezius	—adduction of scapula —fixes inferior angle of scapula to thoracic wall	Weakness	—scapula abduction with outward rotation of inferior angle —winging of inferior border of scapula —increased thoracic kyphosis
Cervical erector spinae	—extension of cervical spine	Tightens	—loss of forward bending —loss of axial extension —holds cervical spine in forward-head posture
Anterior cervical musculature	—flexion of cervical spine	Weakens	—weakness in forward bending —loss of axial extension —inability to pull out of forward-head posture

terior hip structures against the erector spinae as antagonistic groups. This situation is more common in men with early to moderate degenerative joint disease of the lumbar spine. The tightness in the hamstrings and posterior capsule of the hips pulls the spine into forward flexion, holding the erector spinae in a lengthened position, leading to progressive weakness. The corresponding dysfunction is usually joint hypermobility with eventual instability of the lumbar spine (Figure 7–7 and Tables 7–4 and 7–5).

The clinician should consider these myofascial relationships and how they correlate to structure when evaluating posture. These findings may then be correlated to the remainder of the evaluation.

ACTIVE MOVEMENT ANALYSIS

Evaluation of active movements gives the clinician more valuable information regarding possible pathology of the spine or extremities that

Figure 7–10

may be correlated with postural findings. In evaluating active range of motion from a myofascial standpoint, the clinician should first look regionally, then segmentally. Regional observation will usually reveal myofascial abnormalities, whereas segmental observation reveals more specific joint abnormalities. Entire spine motion should be observed, with the patient being instructed to move segmentally starting in the cervical area and proceeding through the thoracic and lumbar spines. Spinal movements

Table 7–4 Lumbar/Lumbopelvic Agonist/ Antagonist

Postural	Phasic
Iliopsoas	Gluteus maximus
Tensor fasciae latae	
Hamstrings	Quadriceps
Hip adductors	Gluteus medius
Gastrocnemius-soleus	Dorsiflexors
Erector spinae	Abdominals
Piriformis	

should be observed in total at least once, regardless of the suspected area of pathology. The area of pathology should then be examined specifically. The reason for performing both regional and segmental observations is that many times, dysfunction that is symptomatic in one area of the body can be caused by a primary dysfunction in another area of the body that is not symptomatic, but needs treatment to resolve the symptomatic dysfunction. This is especially true when examining the myofascial system, because the fascial planes are more regional, as are their dysfunctions.

Restriction of movement in the posterior musculature and fascia of the lower extremity, with corresponding hypermobility of the lumbar spine, exemplifies regional, asymptomatic dysfunction causing symptomatic dysfunction elsewhere. The patient, usually a man, has low-back pain, and with active movements, exhibits an exaggerated lumbar curve reversal. The pelvic contribution to forward bending is limited because of tight hips, hamstrings, and posterior fascial planes. Over time, the posterior structures of the lumbar spine become stretched and hypermobile, creating lumbar instability. The primary dysfunction that needs to be addressed includes the hips, hamstrings, and posterior fascial structures in order to balance the contributions of the hip and low back to overall forward bending. The patient usually has a flattened lumbar lordosis; the loss of lordosis is correlated with regional movement patterns to assess the primary and secondary dysfunctions. Looking only segmentally in the lumbar spine can cause the clinician to miss the primary causative dysfunction.

As with a standard structural examination, all the cardinal plane movements including forward bending, side bending, and rotation should be observed. Quadrant movements should also be observed because daily movements and resulting dysfunctions occur in multiplane dimensions. This again is especially important when dealing with the myofascial system, since it is multidirectional. The multiplane motions that are useful to observe are: (1) forward bending,

Table 7–5 Muscle Agonist/Antagonist Groups of the Lumbopelvic Area and Resulting Dysfunction

Muscle Group	Action	Response to Dysfunction	Results of Dysfunction
Iliopsoas	—hip flexion —assists in external rotation and adduction —backward bending of lumbar spine —anterior ilial rotation	Tightens	—restricted hip extension —tight anterior capsule —increased lumbar lordosis —decreased posterior rotation of ilium
Tensor fasciae latae	—hip flexion, internal rotation, abduction —anterior ilial rotation —knee flexion assistant	Tightens	—restricted hip extension, ER, adduction —decreased posterior rotation of ilium —contributes to increased lumbar lordosis
Gluteus maximus	—hip extension —posterior rotation of ilium	Weakens	—loss of hip extension —decreased posterior rotation of ilium
Hip adductors	—hip adduction —assist in hip flexion —anterior rotation of ilium	Tightens	—restricted hip abduction —restricted posterior rotation of ilium
Gluteus medius	—hip abduction —ant. fibers-IR hip —post. fibers-ER hip	Weakens	—limited hip abduction —loss of lateral stabilization of hip joint
Erector spinae	—extension of spine	Tightens	—increased lumbar lordosis —pelvis tilted anteriorly
Abdominals	—flexion of spine	Weakens	—tendency for pelvis to tilt anteriorly —tendency toward increased lumbar lordosis

side bending, and rotation to the same side; and (2) backward bending, side bending, and rotation to the same side. The first combined set of motions follows a very functional movement pattern that usually helps assess, among other things, the flexibility of the myofascial planes on the contralateral side of the movement. The second combined movement is generally used to assess compressive joint lesions of the spine on the same side the movement is occurring. When the same extension quadrant is observed

from the anterior view, the anterior fascial planes can be evaluated for restrictions. Because the diaphragm and anterior fascial planes may become restricted in the forward-head posture, observing the backward bending quadrant movement from an anterior angle is important.

Compressive Testing of the Spine

Compressive testing of the spine is usually considered a special test of the spine, but should be routinely performed. A convenient time to perform this test is after active movement testing. The concept behind compressive testing is to test the amount of "spring" that the spine has when a direct compression is imparted (Figure 7–11). Generally, patients with accentuated curvatures will have an increased springiness, indicating increased lever arms for the effects of gravity and increased stresses on myofascial structures. The spines of patients with decreased curvatures (axially extended cervical spine along with decreased lordosis in the lumbar spine) will not have enough "spring," leading to decreased shock attenuation during normal everyday activities. Ballistic or impact exercise such as jogging or aerobic exercise may further accentuate the dysfunction. Postural reeducation after normalization of myofascial tone can help correct this dysfunction.

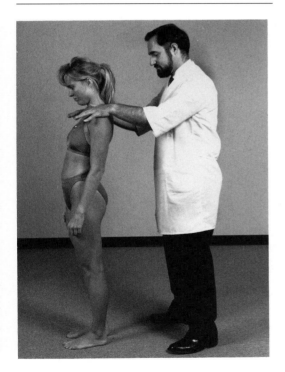

Figure 7–11

PALPATORY EXAMINATION

Once posture and active movements are assessed, the clinician may begin to estimate where the significant dysfunctions exist. The palpatory examination reveals yet more information that may be correlated to previous findings, and offers a clear picture of possible goals and treatment approaches.

The palpatory examination includes, but is not necessarily limited to: (1) palpation of the myofascial structures in the form of layer palpation, (2) palpation of joint structures, and (3) assessment of passive segmental mobility. Palpation of myofascial structures is primarily emphasized here, including layer palpation and passive mobility of muscles and fascial mobility.

Layer Palpation

Layer palpation is a systematic method of assessing the mobility and condition of the myofascial structures, starting from the most superficial structures and progressing into the deepest palpable structures. Layer palpation is extremely important, especially since a common error in both assessment and treatment is to delve into the deeper structures without assessing the superficial structures. The tissues that can be palpated include the skin, subcutaneous fascia, blood vessels, muscle sheaths, muscle bellies, musculotendinous junctions, tendons, deep fascia, ligaments, bone, and joint spaces. The clinician should be able to palpate in depth the location of the structures during the palpatory examination. Is only the skin being palpated or is the subcutaneous fascia also being palpated? Is the muscle sheath being palpated, or has the muscle belly been penetrated? Is the clinician palpating the musculotendinous junction or the tendon itself? Perfecting layer palpation requires development of tactile as well as visual senses. The development of tactile skills includes the ability to detect tissue texture abnormalities. How is the tissue at that level different from surrounding tissues at the same level of depth, or the tissue on the contralateral side?

Table 7–6 Descriptive Terms for Layer Palpatory Exam

superficial—deep	acute—chronic
compressible—rigid	painful—nonpainful
moist—dry	circumscribed—diffuse
soft—hard	rough—smooth
hypermobile—	thick—thin
hypomobile	

For practical purposes, the layer palpation format may be categorized into superficial and deep palpation (Tables 7–6 and 7–7). The superficial palpatory examination includes tissue temperature and moisture and light touch to determine the extensibility and integrity of the superficial connective tissues. Tissue rolling is an important part of layer palpation; it gives the clinician information about the extensibility of the subcutaneous connective tissue (Figure 7–12). In tissue rolling, the skin and superficial connective tissue are lifted up, away from the deeper tissues. The extensibility of the tissues, as well as the integrity of the tissues may be palpated.

The deep palpatory examination includes compression, which is palpation through layers of tissue perpendicular to the tissue, and shear. Shear is movement of the tissues between layers, moving perpendicular to the tissue. The structures palpable are muscle sheaths, muscle bellies, tendons, myotendinous junctions, tenoperiostial junctions, joint capsules, and the deep periosteal layers of tissue. Tissue texture abnormalities and restrictions are noted in this evaluation. Transverse muscle play is an effective assessment tool for assessing the mobility of a muscle or muscle group within the enveloping fascial sheath. The muscle is "bent" in order to assess the transverse flexibility of the muscle. This concept is elaborated on in Chapter 8.

Once the evaluation is completed, the findings are correlated to define the specific dysfunction and treatment is initiated accordingly. Reevaluation is taking place before, during, and after treatment and the treatment is adjusted to accommodate changes being made.

Table 7–7 Palpatory Exam

	Elements of Evaluation	Structures To Palpate
Superficial examination	—Light touch —Tissue temperature and moisture —Mobility of superficial fascia —Skin rolling	—Skin —Superficial connective tissue
Deep examination	—Compression: palpation through layers of tissue perpendicular to the tissue —Shear: movement of tissues between layers perpendicular to tissue	—Muscle sheaths —Muscle bellies —Tendons —Myotendinous junction —Joint capsule —Periosteal layer

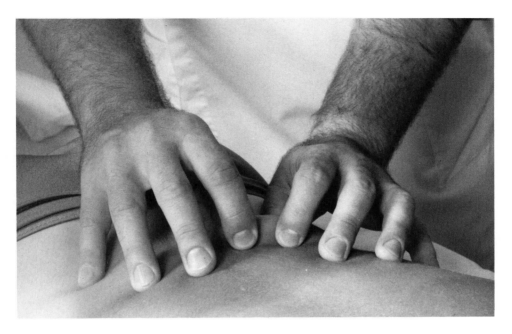

Figure 7–12

REFERENCES

1. *Dorland's Illustrated Medical Dictionary*, 25th ed. Philadelphia: WB Saunders; 1974.

2. Grodin AJ, Cantu R. *Myofascial Manipulation*. St. Augustine, FL: Institute of Graduate Physical Therapy; Course notes.

3. Cyriax J. *Textbook of Orthopaedic Medicine: Diagnosis of Soft Tissue Lesions*. London, England: Baillière Tindall; 1:46.

4. Steindler A. *Kinesiology*. Springfield, IL: Charles C Thomas; 1977:35–37.

5. Goldenherg DL. Fibromyalgia syndrome: an emerging but controversial condition. *JAMA*. 257:2782–2803.

6. Janda V. Muscles, central nervous motor regulation and back programs. In: Korr I, ed. *The Neurobiologic Mechanisms in Manipulative Therapy*. New York: Plenum; 1978:27–42.

CHAPTER 8

Atlas of Therapeutic Techniques

Robert I. Cantu and Alan J. Grodin

The following atlas of therapeutic techniques is by no means a comprehensive treatment of all myofascial technique. It merely represents a compilation of techniques that, in the opinion of the authors, have consistently proven to be effective in the clinic. The purpose of the book, and specifically of this chapter, is to give the clinician a solid and basic understanding of myofascial technique. As the techniques are used, the clinician will modify them to meet the individual needs of both patient and clinician. The techniques then become personalized, and therefore, unique to that particular practitioner. New techniques are born in this way and, many times, evolve into specific systems of treatment. Myofascial manipulation has undoubtedly been performed since the beginning of time, and has evolved into its present-day variety of formats. Myofascial manipulation will continue to evolve into more effective applications as the body of knowledge increases.

Before discussing individual technique, certain terms should be defined and treatment concepts and procedures discussed, for the sake of clarity and consistency throughout the chapter.

Joint versus soft tissue manipulation: Some difficulty may arise in drawing the line between what is soft tissue manipulation and what is joint manipulation. If a *joint* is operationally defined as "a space built for motion," then any tissue surrounding the "joint" may be considered

soft tissue. Ligament, capsule, periosteum, and fascia are all histologically classified as connective tissue. When dealing with the joint, the following concept may be applied: Anything that is not bone is connective tissue. Technically speaking, then, joint mobilization is a form of soft tissue mobilization since the extensibility of the connective tissue surrounding the joint is being changed.

For the purpose of clarity in this text, however, the operational definition of a joint should be expanded. A joint may be defined as "a space built for motion in which movement is governed by (a) arthrokinematic rules and (b) connective tissue extensibility." The arthrokinematics is the distinguishing factor in separating soft tissue mobilization from joint mobilization. Joint restrictions occur and are treated in characteristic arthrokinematic fashion. Mobilization technique must be applied following arthrokinematic rules in order to restore extensibility. Myofascial restrictions, on the other hand, are not as predictable since they can occur outside the realm of specific joint arthrokinematics. Restrictions of the superficial fascia, for example, may occur in many planes and in many different—and unpredictable—directions. The treatment is based on localizing the restriction and moving into the direction of restriction, whether or not the direction follows the arthrokinematics of the nearby joint.

Herein lies one of the problems with myofascial manipulation: Treatment has a tendency to become subjective and abstract. The danger of losing credibility is higher than in joint manipulation, since treatment is based on "what the therapist is feeling." There is no doubt that "good hands" and an "intuitive mind" are of great value in manual therapy, specifically in myofascial manipulation. A balance should exist, however, between scientific scrutiny and clinical intuition. Treatment that relies heavily on one while de-emphasizing the other will not be balanced, and, therefore, not be as effective. This text represents myofascial manipulation in a biomechanical and kinesiological sense, respecting and integrating nearby joint arthrokinematics as much as possible. In this way, myofascial manipulation is represented in the most concrete empirical form possible, without negating the intuitive aspects of the treatment technique.

Sequencing of treatment: The sequence in which technique is applied will generally spell the difference between success and failure. The question is: Where in the entire treatment scheme does myofascial manipulation fit? And how does the clinician sequence individual myofascial technique for optimal results? Each patient is different and each clinician will determine the sequence of treatment on an individual basis; however, the guidelines discussed below may be helpful in deciding treatment sequencing for individual patients. A general scheme of treatment is as follows.

1. Myofascial manipulation of involved and regional areas associated with local involvement. With joint mobilization, the treatment often focuses on individual joints being moved in specific directions. Myofascial manipulation, however, generally focuses on larger areas or regions of treatment. Individual joint restrictions often have significant myofascial components. Passive segmental mobility of individual joints may change with regional treatment of myofascia. Releasing myofascial tissues prior to joint mobilization also allows joint mobilization and/or manipulation to be performed with less force application. If the myofascial component

of the restriction is first released, the mechanical restriction can more easily and more specifically be treated. The general progression of myofascial manipulation considers the following factors:

a. Direct before indirect technique. For the most part, all the techniques described in the text are direct ones. In other words, the techniques locate the restriction and move into the direction of the restriction. If the changes cannot be made with direct technique—because of pain, autonomic responses, or severity of the restriction—indirect technique may be used. The concept is that the shortest distance between any two points is a straight line, and the shortest distance through a restriction is directly through the restriction.

b. Superficial to deep. Common sense dictates that application of myofascial technique begins superficially and progresses into depth as changes are made, or in search of deeper myofascial restrictions. Treatment that progresses from superficial to deep also allows the patient gradually to grow accustomed to the clinician's hands; this facilitates relaxation and allows for unforced penetration to deeper levels. Deeper technique is not synonymous with more aggressive technique. If the deeper connective tissues are properly accessed, they may be treated effectively without potential microtrauma and exacerbation of symptoms. Instead of breaking down the doors, the clinician allows the body to open the doors for easy and less damaging access into an area.

2. Joint mobilization after treatment of myofascia. As the myofascia releases, joint mobilization becomes easier, and individual joints are more easily isolated. At times, however, if the myofascial restriction is unyielding, joint mobilization and/or manipulation may become necessary to free up the myofascia. The type III joint mechanoreceptors, which are stimulated by joint manipulation, inhibit surrounding muscular activity. Joint and myofascial manipulation are "played off" one another—joint mobilization inhibits myofascia, and myofascial manipulation facilitates joint manipulation.

3. Joint and myofascial elongation. Once extensibility has been improved in the myofascia and the joints, elongation and stretching may be approached with greater efficiency. Elongation (distinct from stretching) refers primarily to the spine, where the forces applied "open the accordion" and decompress the spine. No specific stretch is applied but decompression forces are applied. In the lower extremity, for example, myofascial manipulation should always be performed on a hamstring prior to stretch to allow for greater tissue extensibility.

4. Neuromuscular reeducation. Stretching and strengthening exercises and movement approaches (i.e., Feldenkrais,[1] Alexander[2]) are appropriate at this time. The alternate somatic movement therapies correspond with the concepts of myofascial and joint manipulation, but their effectiveness is limited if the tissue is not first prepared. The new extensibility of the tissue obtained from the myofascial and joint manipulation, the stretching, and elongation facilitates the promotion of new movement patterns. Patients are encouraged at this time to stretch, strengthen, and move in new, more efficient patterns.

5. Postural instruction. Once the restrictions are removed and the patient freely moves in new, more efficient patterns, the potential exists for postural reeducation. If postural instruction, which is necessary for most patients, is given at the beginning of the treatment sequence, the patient cannot effectively assume the new postures. The patient tires from moving against his or her own restrictions, and a negative feedback loop is established. The patient reports, for example, that "it is easier to slump than to try to sit erect," and the poor postural pattern is actually reinforced. With new freedom of movement, good posture is easier and is positively reinforced.

Positioning of patient and therapist: To achieve maximal therapeutic effect, both patient and therapist should be situated in the most efficient positions possible. This concept may seem elementary, yet it is often forgotten in the day-to-day treatment of patients. Any inefficiency in the therapist's application of the treatment is transferred to the patient. The patient senses this inefficiency in the manual technique and is unable to relax fully.

The second aspect of positioning, included in the discussion of specific techniques, is the generous use of pillows, especially between the patient and the therapist. *When positioning patients, especially in the sidelying position, a pillow should always be placed between the therapist and the patient.* The pillow provides a mechanical barrier between patient and therapist, which aids the biomechanical delivery of the technique and avoids needless body contact.

1. The use of body weight, ground, and lever arms. Since physical longevity is important to the manual therapist, and since many times the patient may outsize the therapist, the use of body weight, ground, and lever arms is important. Use of body weight can be optimized by utilization of high-low tables, or by the therapist standing on a stool or step. The ability to lean over the patient acts as a significant force multiplier, whether the patient is prone, sidelying, or supine.

2. The use of the ground is all about weight shifting. When applying technique in a push/pull type of technique, weight shifting allows the therapist to access the lower kinetic chain. Rather than being "all arm," the use of the lower kinetic chain also becomes a force multiplier. The hands also become more relaxed in the application of the technique, and the technique becomes more forceful, but softer at the same time.

3. Using lever arms whenever possible is yet another force multiplier available to the therapist. The longer the lever arm, the greater and more focused the force becomes. This is especially important in joint mobilization, but is applicable to soft tissue mobilization as well. A precautionary note is in order at this point. The longer the lever arm, the greater the force multiplication, the greater the risk of injury. Some manual therapists advocate the use of shorter lever arms for greater safety, and their point is well taken. The manual therapist should be careful when lengthening the lever arm, recognizing the force multiplication that is occurring.

Care and protection of hands: The hands are the primary treatment modality for the manual therapist and do not come with a replacement guarantee. If a manual therapist sees 15 patients a day 5 days per week for manual therapy, the therapist is laying hands on more than 3700 bodies per year. The numbers accumulate during the course of a career. The hands are very durable body parts; however, the principles of Wolf's law (good stress/bad stress) all apply. Practicing correct application of technique and following proper hand-care procedures are essential for ensuring longevity of the manual therapist's career. The following are some suggestions for hand care:

1. Whenever possible, use techniques that do not hyperflex or hyperextend any joints (Figure 8–1). End-range maneuvers will only accelerate joint hypermobility problems, leading to early arthritic changes. The thumbs should be aligned with the metacarpals, which in turn, should be aligned with the radius. The thumb and proximal interphalangeal joint (PIP) of the index finger can be used together to form a very stable contact surface (Figure 8–2).

2. Adapt for therapist/patient size differences. If the patient is large-sized, and the desired depth of penetration is not practical, do not use the fingers or thumb. The fist and elbows are excellent alternatives. Palpate with the fingers; treat with the elbows or fist.

3. Wash hands in cold water after each patient treatment. If any inflammation occurs during a patient treatment, the cold water may act as a cryotherapy/anti-inflammatory treatment. Warm water 15 to 20 times per day may have a cumulative inflammatory effect, whereas the cold water may slow the process down.

4. Protect the hands during off hours. When gardening or performing any type of work that may be hard on the hands, the therapist should wear gloves. Manual therapists actually incur

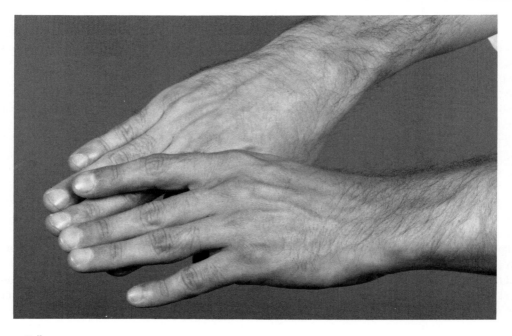

Figure 8–1

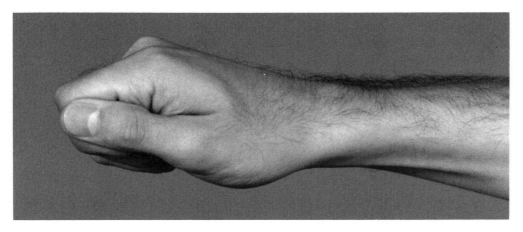

Figure 8–2

more microtrauma to their hands in the off hours, during the time when the hands should be getting much-needed rest.

5. Use of lubricant. A small amount of lubricant should be used, especially in techniques involving longer stroking. The amount of lubricant should be just enough to decrease noxious skin friction, but not enough to cause slipping of the hand on the body. A certain amount of "traction" on the skin is necessary for appropriate delivery of the technique.

TECHNIQUES FOR THE LUMBAR SPINE

Bindegwebbsmassage-Type Stroke (Figures 8–3 to 8–6)[3]

Purpose: This technique is a reflexive or autonomic technique; it is used when the patient shows signs of being autonomically facilitated or extremely hypersensitive. Many patients exhibit acute symptoms that mimic a reflex sympathetic dystrophy. The skin, for example, is hypersensitive with a cold clammy feel or touch, and the patient is easily nauseated. For a patient

with such symptoms, starting with a deep touch is usually counterproductive. The technique suggested here offers an entry way into deeper technique by quieting the autonomic system.

Patient position: Prone.

Therapist position: The therapist stands over the patient, perpendicular to the patient.

Hands: Contact will be made with the pads or tips of the last 3 fingers. The pisiform is the axis of motion for the technique.

Execution: One hand will be placed on the patient to stabilize gently the subcutaneous connective tissue. The treatment hand is placed gently on the patient, with the pisiform being the axis of motion for the technique. Starting with the elbow close to the body, the elbow is moved away from the body, bringing the fingers away from the stabilizing hand. The technique is repeated at a deliberate pace, moving about an area of the spine as indicated. The technique is superficial, going only as deep as the superficial, subcutaneous connective tissue. The technique is generally comfortable, and at worst, should be only mildly uncomfortable. Remember, the goal of this technique is to quiet the autonomic system, not to create mechanical changes.

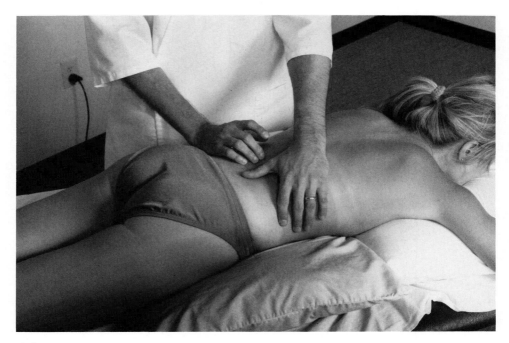

Figure 8–3

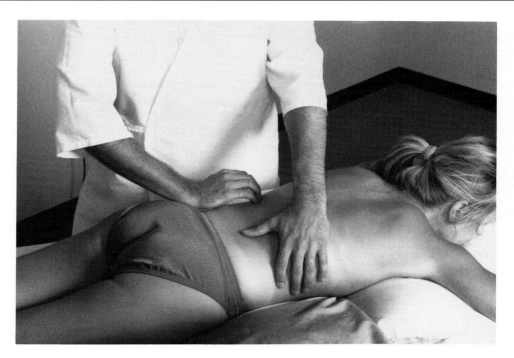

Figure 8–4

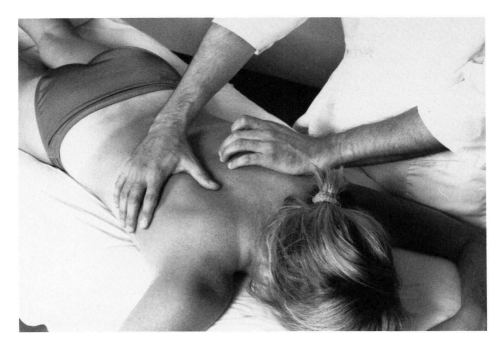

Figure 8–5

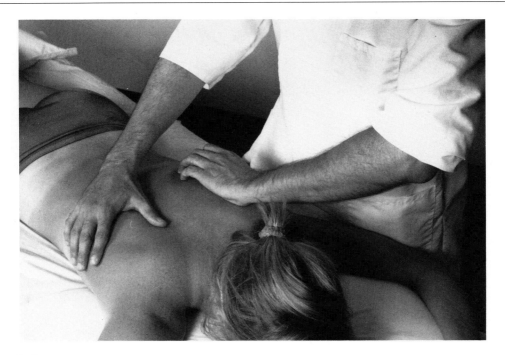

Figure 8–6

Long Axis Distraction of Superficial Connective Tissue (Figures 8–7 and 8–8)

Purpose: The purpose of this technique is elongation of the superficial connective tissues, usually in the cephalad-caudad direction. Since the subcutaneous connective tissue is multidirectional in the fiber orientation, diagonal restrictions may occur and should be treated. This technique can also be performed on a deeper level to provide an elongation of the spine itself.

Patient position: The prone position is demonstrated here, but the technique can be performed in any position depending on the location of the restriction. In the supine position, for example, the technique can be used to treat restrictions in the anterior chest or abdomen.

Therapist position: The therapist stands over the patient, perpendicular to the direction of the restriction.

Hands: Hands are placed in a crossed position on the patient, directly in line with the restriction.

Execution: The therapist applies gentle anterior pressure until the subcutaneous fascial level is reached. A gentle distraction is then applied in the direction of the restriction, usually cephalocaudal. The technique can be performed in the midline, off-center, diagonally, or in any direction of restriction. When being performed in the midline with a deeper pressure, a distraction and elongation of the spine will result. Care must be exercised with the deeper version of the technique in patients with degenerative joint disease or discogenic lesions.

Medial-Lateral Fascial Elongation (Figures 8–9 and 8–10)

Purpose: The purpose of this technique is to elongate the superficial fascia in a medial-lateral direction. As with the previous technique, the application may be superficial as well as deep. The most superficial application of the technique is autonomic, whereas any deeper application is primarily mechanical.

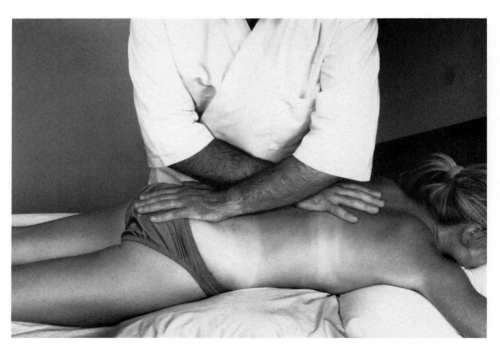

Figure 8–7

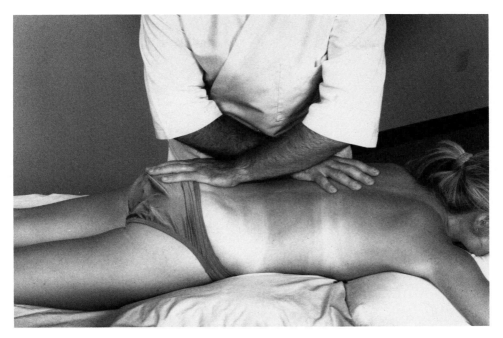

Figure 8–8

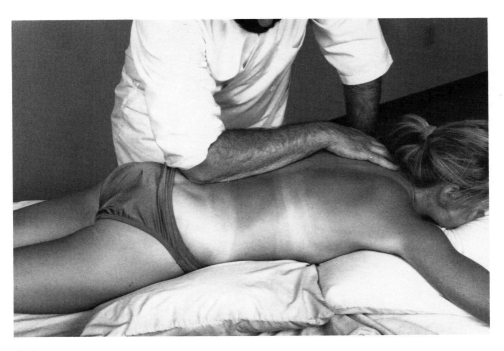

Figure 8–9

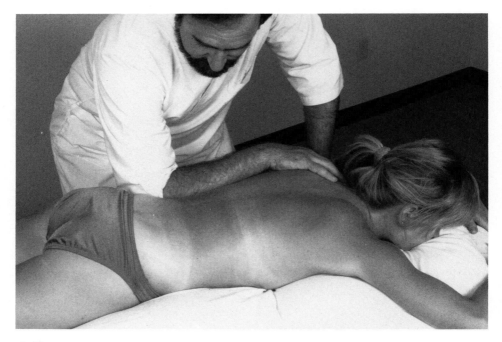

Figure 8–10

Patient position: Prone.

Therapist position: The therapist stands perpendicular to the patient, with the top hand on the treatment table for support and efficiency in application of technique. The other elbow is placed in the area of the lumbosacral junction with the forearm and hand resting lightly on the patient.

Execution: The therapist applies gentle anterior pressure with the elbow until the level of superficial subcutaneous fascia is reached. Lateral elongation pressure is then applied, and the elbow is allowed to slide laterally and around the body. Most of the pressure is at the elbow and the proximal one third of the ulna. The rest of the forearm is merely resting on the patient as the technique is executed. As the subcutaneous fascia releases, and as patient tolerance dictates, deeper pressure may be gradually applied to the muscular and periosteal levels.

Tissue Rolling (Figures 8–11 and 8–12)

Purpose: The purpose of this technique is mechanical assessment and alteration of restrictions in the superficial fascia.

Patient position: Prone.

Therapist position: The therapist stands diagonally over the patient.

Execution: Assessment: The skin and subcutaneous fascia are gently lifted in a posterior direction at different levels and areas of the spine. Generally, the tissue is assessed just off the midline of the spine and in a caudal to cephalic direction. Typically, the fascia directly over the spine has much less mobility; this decrease should not be considered dysfunctional.

As with other superficial techniques, the assessment may be in medial-lateral or diagonal directions because of the multidirectionality of the superficial connective tissue. Some patients

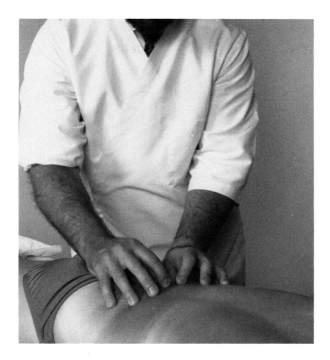

Figure 8–11

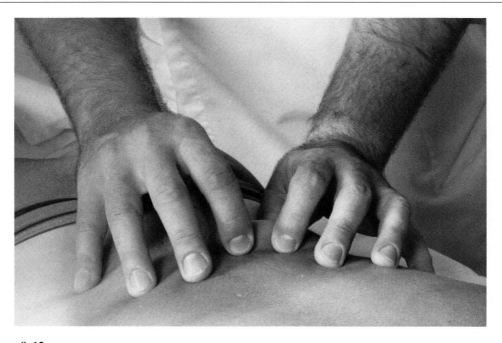

Figure 8–12

will be quite restricted in all planes; this may be a general function of body type, or may represent generalized restrictions. The clinician must not only base the clinical judgment on the superficial fascial assessment, but must also correlate the findings with other components of the evaluation.

Therapeutic application: The skin and superficial subcutaneous connective tissue are gently lifted in a posterior direction with both hands. Using each hand alternately, the clinician rolls the skin, never releasing the hold on the skin and subcutaneous tissue. Generally, the skin is rolled from caudal to cephalic, but other directions such as medial to lateral or diagonals can be pursued. One can imagine balancing a drop of water on the lifted portion of the skin as the roll is applied. When a restriction is encountered, the rolling can be stopped, and a gentle posterior stretch or oscillation can be applied.

Long Axis Laminar Release (Figures 8–13 and 8–14)

Purpose: The first purpose of this technique is elongation and decompression of the spine. The second purpose is the identification of localized lesions in the medial border of the erector spinae. As these lesions are identified, the motion may be stopped and a sustained pressure may be applied.

Patient position: The patient is positioned prone with the lumbar spine in a neutral position. The neck also should preferably be in a neutral position and not rotated. The patient's head should be as close as possible to the head of the table to allow the therapist to complete the technique through the iliac area.

Therapist position: The therapist is positioned at the head of the table with one foot in front of the other.

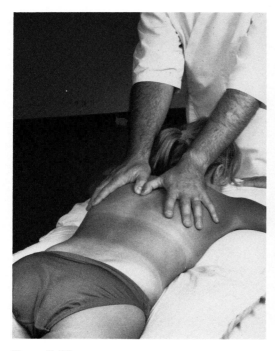

Figure 8–13

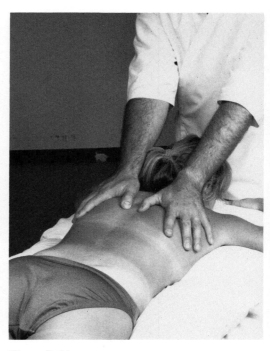

Figure 8–14

Hands: The hands are placed gently over the patient with the fingers and thumbs facing in a caudal direction. The thumbs are placed in the groove between the erector spinae and the spine. The technique is best performed with both thumbs on a single side of the spine, one thumb just behind the other. A bilateral technique can also be performed but the depth of penetration is somewhat compromised. Note that the thumbs should be aligned so they are in a direct line with the radius. This alignment allows for the most efficient application of technique and the least amount of biomechanical compromise for the therapist's hands.

Execution: Starting in the upper thoracic area and with moderate pressure in the groove between the erector spinae and the spine, the thumbs are moved caudally into the lumbar and lumbosacral areas. As the lumbosacral junction is reached, the palms of the hands engage the iliac crests, and a gentle traction force is applied. After several strokes, lesions along the groove may be identified. These lesions are manifested as local increases in muscle tone, reflexive muscle guarding, or connective tissue thickenings. The lesions may be results of acute inflammation or may be remnants of older trauma, holding patterns, or chronic fibrotic changes. The movement of the hands may be stopped at any time to apply localized sustained pressure.

Muscle Play of Erector Spinae (Figures 8–15 and 8–16A, B)

Purpose: This technique mobilizes the fascial sheath or casing surrounding the erector spinae. As previously defined, muscle play is "the ability of the muscle to expand and move within its compartment independent of joint movement

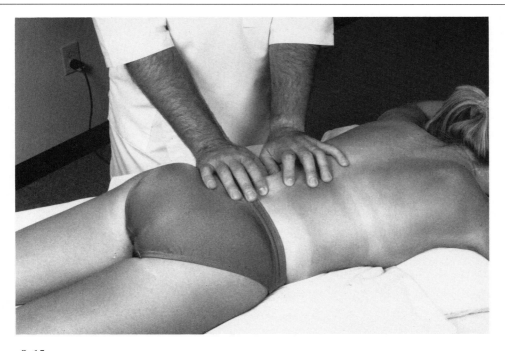

Figure 8–15

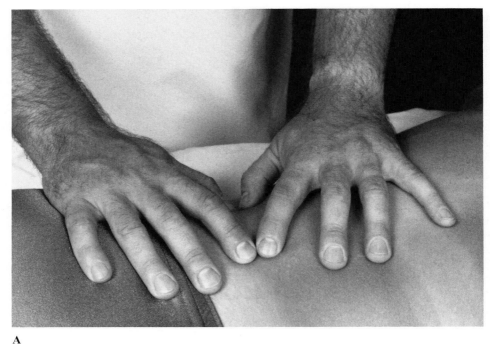

A

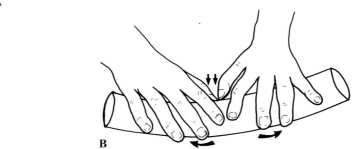

B

Figure 8–16

or voluntary muscle contraction." Many fascial restrictions occur in planes perpendicular to or diagonal to the direction of the muscle fibers. Recall that muscle sheaths are classified as loose connective tissue that has multidirectional fiber orientation. By mobilizing the connective tissue sheath surrounding or encasing the muscle or groups of muscle, muscular contraction can occur more efficiently, circulation to the muscle is improved, and movement in the localized and general areas is improved.

Patient position: Prone.

Therapist position: The therapist is standing perpendicular to the patient.

Hands: Hand position for this technique is extremely important. The movement can be likened to the bending of a garden hose. If one imagines a garden hose being an encasement in which improvement of mobility is desired, bending the hose is one way to accomplish this goal. For the technique, the thumbs are placed on the lateral border of the erector spinae. Once again, the thumbs should be positioned so they are in line with the radius of the forearms. This ensures that forces are distributed throughout the arm and are not localized in the interphalangeal (IP), metacarpophalangeal (MCP), or carpometacarpal (CMC) joints. Having the thumbs in any other position will quickly produce fatigue. The index fingers are placed lightly over the medial border of the erector spinae. The palms of the hands are resting lightly over the lateral aspect of the patient's body (Figure 8–16A, B).

Execution: This technique is performed in an oscillatory manner using a medial-lateral direction of force. Initially, the force is applied through the palms, allowing the patient's body to oscillate segmentally, primarily in a medial to lateral direction. This rhythm will vary from patient to patient and will also depend on the patient's general state of relaxation. Moving too quickly or slowly will result in either a logrolling type of motion or a motion that is out of resonance. Once a satisfactory rhythm and excursion are attained, the thumbs, which are contacting the lateral border of the erector spinae, begin to create the bending force in synchrony with the rhythm of the rest of the body. The primary force is now at the thumbs, with the palms retaining a degree of force to maintain the oscillation. The "power" portion of the stroke is lateral to medial with the thumbs; the index fingers are merely monitoring the position of the hand on the erector spinae. To ensure that a bending movement is being executed (as opposed to only a medial-lateral movement), the elbows must move from a position away from the body to a position toward the body during the power portion of the stroke. In other words, the elbows are held away from the body at the initiation of the stroke (shoulder abduction) and are moved toward the body during the stroke (shoulder adduction).

If a restriction is identified in a medial to lateral direction, the hand position is changed or reversed, so the thumbs are contacting the medial border of the erector spinae. The therapist must, therefore, move to the other side of the table to perform the technique. The "power" portion of the stroke is still delivered through the thumbs, but now in a medial to lateral direction. Different levels of the erector spinae may be treated by simply moving the hands cephalic or caudal, being sure that the thumbs contact the lateral borders of the erector spinae.

"Ironing" of Erector Spinae Muscle Group (Figure 8–17)

Purpose: The purpose of this technique is to produce tonal inhibition of the erector spinae muscle group while applying gentle unilateral traction to the lumbar spine. Since longitudinal manipulation is usually less noxious and more sedative than cross fiber manipulation, this is an excellent technique for applying moderately deep pressure when the patient is in considerable discomfort or pain.

Patient position: Prone.

Therapist position: The therapist stands directly facing the patient at the approximate level of the lumbosacral area.

Hands: The top hand is placed over the iliac crest to "anchor" the pelvis. The bottom hand is crossed over the top hand and placed over the erector spinae muscle mass as close to the lumbosacral junction as possible. The table should be low to allow for the use of the therapist's body weight.

Execution: A small amount of lubrication is used. The palm of the bottom hand pushes into the erector spinae muscle group and slides slowly and firmly in a cephalic direction. This technique is deep, but utilizes the entire heel of the hand to create a strong but diffuse technique. During the technique, the top hand remains anchored onto the iliac crest, allowing for a moderate traction/distraction of the lumbar area.

Bony Clearing of the Iliac Crest (Figures 8–18 and 8–19)

Purpose: This technique is designed to first evaluate the fascial attachments at the iliac crest, then soften the fascia especially at the insertion of the deep erector spinae and quadratus lumborum. This technique also serves to prepare the

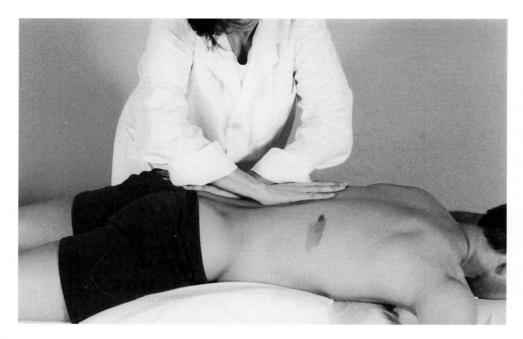

Figure 8–17

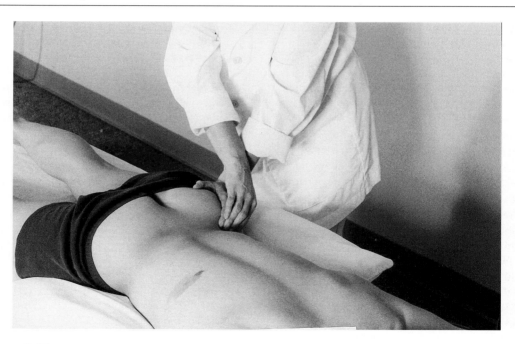

Figure 8–18

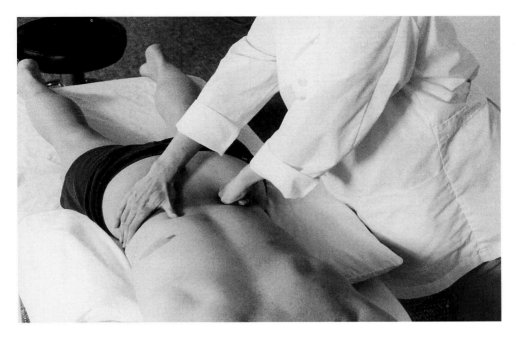

Figure 8–19

iliac crest surface area for the next series of techniques (iliac crest release).

Patient position: Prone.

Execution: In the first part of the technique, the fingers of both hands are placed directly over the superior border of the iliac crest. For better mechanical advantage, the fingers of one hand are placed over the fingers of the hand making contact with the patient. The technique starts on the superior border of the iliac crest, as close to the midline as possible. The fingers scour along the superior border of the iliac crest laterally and at moderate depth. A small amount of lubricant should be used to avoid overly frictioning the skin.

In the second part of the technique, the "power grip" shown in Figure 8–2 is utilized to gain further depth. Again starting as medially as possible, the therapist scours along the superior border of the iliac crest using the reinforced thumb and PIP joints as the contact on the patient.

Iliac Crest/Lateral Sacral Release (Figures 8–20-A, B, 8–21, and 8–22)

Purpose: This technique mobilizes the fascial planes in the area of the iliac crest and the top one third of the ilium and the lateral border of the sacrum. As previously discussed, the area of the iliac crests contains connective tissue thickenings from various muscular and fascial attachments, and is vulnerable to myofascial restrictions. Movement restrictions in forward bending, side bending, and also backward bending can occur here. The posterior portions of the fascial planes create the forward bending restrictions, whereas the anterior portions create backward bending restrictions.

The lateral border of the sacrum can also be fascially compromised. The piriformis attaches close by, and patients with low back, hip, sacroiliac, and leg pain can profit from this technique. Especially patients with diffuse hip and leg pain proximal to the knee can benefit from this tech-

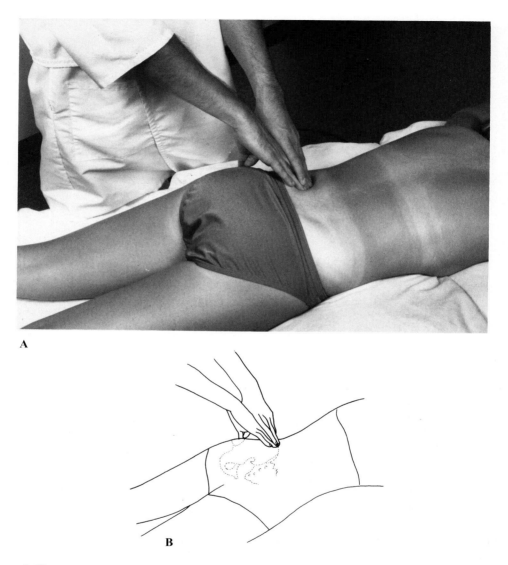

A

B

Figure 8–20

---

nique. The lateral sacral release is an excellent technique to use in conjunction with the bilateral sacral release technique shown next.

Patient position: Prone. Should the connective tissue need to be placed in a slackened position for deeper penetration, the hip may be extended manually by the therapist, or statically with pillows (Figure 8–18).

Therapist position: The therapist stands diagonally over the patient, approximately perpendicular to the iliac crest.

Hands: The optimal hand position for this technique is to have the middle fingers approximating one another (Figure 8–1). The index fingers are "dummy" fingers, one being below and one being above the middle fingers. This posi-

Figure 8–21

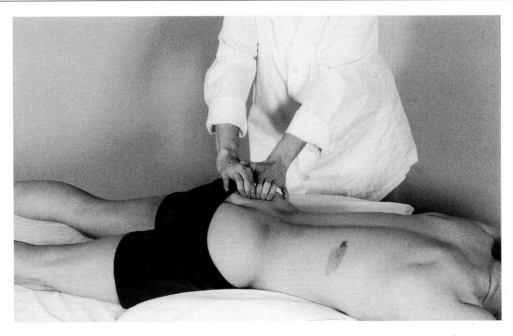

Figure 8–22

tion allows for a four-finger contact on the iliac crest or lateral border of the sacrum.

Execution: The fingers are placed over the border of the iliac crest and an anterior force is applied through the fingers. A very slight extension of the fingers occurs during the power portion of the stroke. The power for the motion comes from the shoulders and upper body and the stroke is delivered repetitively in an oscillatory manner. In correctly applying the force, the fingers will slide off the border of the ilium into the connective tissue. When the fingers are withdrawn posteriorly in preparation for the next stroke, they move back on the border of the ilium. Contact with the patient is never broken during the repetitive application of the technique, except to move to other areas of the iliac crest. The crest may and should be mobilized from the most lateral palpable aspect to the most medial palpable aspect, since the entire border of the iliac crest is susceptible and vulnerable to myofascial restrictions. The depth of penetration of the stroke is moderate and depends on patient tolerance. Many patients with fibromyalgia will be extremely sensitive over this area, whereas many patients will be restricted without experiencing any tenderness. The clinician should treat this area based on objective findings in the evaluation and not merely on subjective complaints. A variation of this technique is to apply the same force, but contact 1 or 2 inches distal to the border of the ilium. As the force is applied over the connective tissue of the ilium, the fingers do not slide off the ilium into the deeper connective tissues. Again, the entire expanse of the ilium should be mobilized, or at least palpated for restrictions.

The same technique is utilized for the lateral border of the sacrum. The fingers start just off the sacrum and push onto the lateral surface of the sacrum in a rhythmical fashion. Remember that the sacroiliac joint occupies the cephalic half of the sacrum. When moving from the ilium to the sacrum on this technique, the therapist "detours" onto the lateral aspect of the posterior superior iliac spine (PSIS) and moves caudally toward the inferior-lateral angles of the sacrum.

The technique covers the bony surfaces starting just lateral to the anterior superior iliac spine (ASIS) and progressing medially and caudally to the sacrococcygeal junction.

Bilateral Sacral Release (Figure 8–23)

Purpose: The purpose of this technique is to mobilize the connective tissue on the sacral borders. This may become necessary before attempting to mobilize the sacrum out of various positional faults or movement dysfunctions. Freeing up the myofascial restrictions often facilitates mobilization of the sacrum. This area may also be restricted in conjunction with iliac crest restrictions. As previously noted, the fascia lata has its insertion at the ASIS, lateral border of the iliac crest, lateral borders of the sacrum, coccyx, and sacrotuberous ligament. To fully mobilize the insertion of the fascia lata, the lateral border of the sacrum should be mobilized.

Patient position: Prone.

Therapist position: The therapist stands perpendicular to the patient.

Hands: The hands are brought together so that the thumbs and the index fingers of each hand are making contact with one another.

Execution: Anatomically, only the distal half of the sacral borders are palpable. The proximal one half of the sacrum articulates with the ilium and is not palpable. To ensure that contact is being made on the sacrum, the therapist should approach the sacrum with the bottom hand below the level of the sacrum (distal to the sacrum), until contact is made bilaterally with the patient's buttock. The bottom hand then palpates in a cephalic direction until the inferior lateral angles of the sacrum are palpated. The top hand then contacts the bottom hand in the manner described above. A repetitive caudal to cephalic motion is performed following the lateral border of the sacrum. The direction of the technique should be V-shaped, following the shape of the sacrum. If the fingers are only moving cephalically and not spreading, contact with the lateral borders of the sacrum is not being maintained.

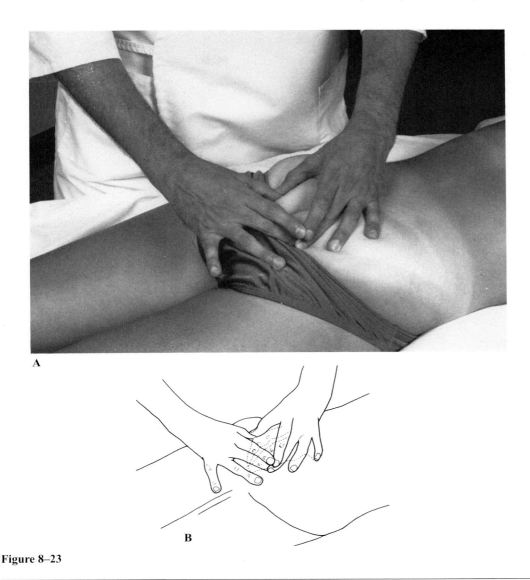

A

B

Figure 8–23

The technique may also be executed unilaterally using the same hand position as the iliac crest release described previously (Figure 8–1). The lateral border of the sacrum is located the same way as described above. Once the lateral border is located, contact is made with the fingertips. The fingers are then moved caudal to cephalic, maintaining contact on the lateral border of the sacrum.

Medial-Lateral Pull Away (Figure 8–24)

Purpose: The first purpose of this technique is autonomic or reflexive in nature. As with other autonomic techniques, it desensitizes the patient who is extremely acute and gains entryway to deeper technique. As the patient's condition allows or dictates, deeper pressure is applied until the level of the erector spinae is reached,

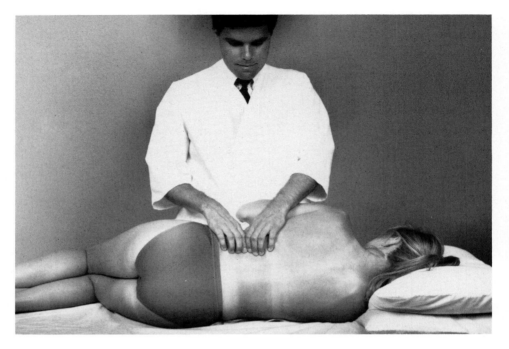

Figure 8–24

changing the emphasis of the technique from autonomic to mechanical. The erector spinae is gently being mobilized from a medial to lateral direction.

Patient position: Sidelying. The patient's hips and knees are semiflexed. As discussed earlier, a pillow should always be placed between the patient and the therapist both for biomechanical advantage and for modesty. The patient is moved close to the edge of the table until snug against the pillow.

Therapist position: The therapist stands over the patient snug against the pillow.

Hands: The hands are placed gently over the patient with the fingertips resting over the medial border of the lumbar erector spinae.

Execution: The stroke begins very gently at approximately the level of the subcutaneous fascia, and from medial to lateral. Initially, the pressure is evenly distributed throughout the hand. As the patient tolerates, more pressure is

exerted through the fingertips until a moderate to deep pressure is being consistently exerted.

L3 (Figure 8–25)

Purpose: The purpose of this technique is to alter the connective tissue in the midlumbar area, and specifically around the L3 area. Since L3 is generally the apex of the lumbar curve, and site of hypomobility problems, myofascial preparation of the area is necessary prior to joint mobilization and/or manipulation. Also, the transverse process of L3 is the longest and most easily palpated.

Patient position: Sidelying. The patient is positioned with the hips and knees in a semiflexed position, and a pillow is placed between the therapist and the patient.

Therapist position: The therapist stands over the patient with the patient snug against the pillow.

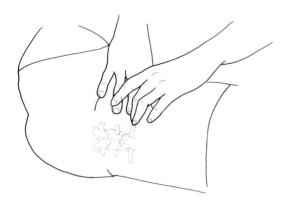

Figure 8–25

Hands: The middle fingertips are used for this technique.

Execution: Starting laterally, the transverse process of L3 is palpated. Once on the transverse process, the fingers are moved superiorly, pos- teriorly, inferiorly, and anteriorly to contact the connective tissue surrounding the L3 transverse process. Once off the transverse process, firm pressure, depending on patient tolerance, is ap- plied with an oscillatory motion.

Passive segmental mobility may be tested in any plane just before and just after the technique is applied. Because soft tissue and joint mobi- lization are often used together, and because joint restrictions may often be due to soft tissue restrictions, passive segmental mobility may be altered with this or any other myofascial tech- nique.

Quadratus Lateral Erector Spinae Release (Figures 8–26 and 8–27)

Purpose: The purpose of this technique is to prepare the quadratus lumborum and the lateral fascial structures of the lumbar spine for elonga- tion and stretch techniques. The technique in- volves sustained pressure primarily designed to

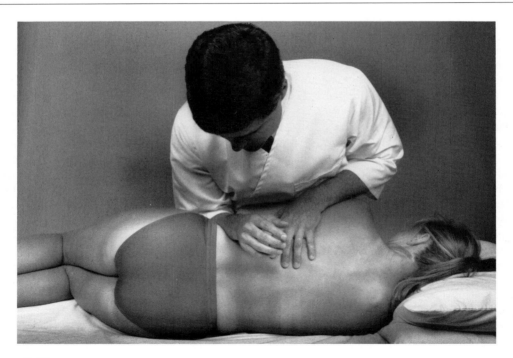

Figure 8–26

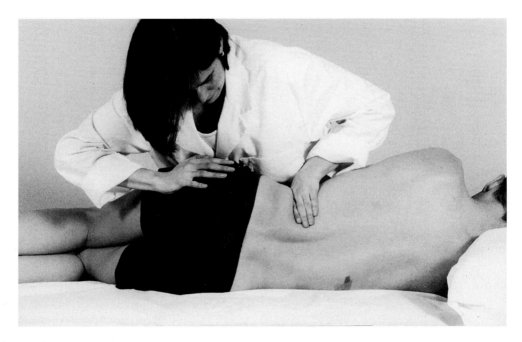

Figure 8–27

reduce active tonic contractions of the quadratus lumborum, and to prepare for a stretch of the lateral fascial structures. After quadratus tone is diminished, the elongation and stretch techniques are more effective and efficient.

Patient position: Sidelying with the hips and knees in approximately 70 degrees of flexion.

Therapist position: The therapist stands perpendicular over the patient. If a high-low table is available, the table level should be lowered.

Hands: The mid forearm of the bottom arm is used in this technique. The forearm is placed in the midlumbar area, in the soft tissue space between the 12th rib and the iliac crest. If the forearm is angled posteriorly, the lateral border of the erector spinae will be contacted. If the forearm is angled anteriorly, the quadratus lumborum will be contacted. As an alternate position, the web space and MCP joint of the top hand can be placed on the quadratus lumborum as the bottom hand positions to hike the hip.

Execution: The top hand is either placed gently on the patient, or on the treatment table for support. The middle aspect of the forearm (ulnar surface) is wedged into the groove between the 12th rib and the iliac crest. Light to moderate pressure is placed down onto the muscle groups and sustained for a period of time until a release of muscular tone is achieved or until it is obvious no change will be made. The forearm may be moved forward and backward (the therapist is flexing and extending the shoulder) in a very deliberate "sawing" type of motion.

As an alternate technique, the hip is hiked using the bottom hand while the quadratus is accessed with the top hand. The top hand is positioned with the first MCP making contact with the quadratus lumborum. As the quadratus is put on slack, the top hand pushes firmly in a medial direction to access the deeper fibers of the quadratus lumborum.

Side Bending Elongation Quadratus Stretch (Figures 8–28, 8–29, and 8–30)

Purpose: This technique should be used generally to elongate the posterolateral and anterolateral fasciae of the lumbar and thoracic spines and, specifically, to stretch the quadratus lumborum. In unilateral chronic pain conditions, the painful side often retracts, contracts, and generally shortens. The manifestation of such a condition can be assessed posturally or with active movements. Both the connective tissues as well as contractile tissues may become dysfunctional and exhibit changes consistent with immobilization.

More specifically, this technique may be used to prepare for correction of lateral shift conditions of more than 3 weeks' duration. As discussed in Chapter 3, muscle decreases in length by losing sarcomeres—the process takes approximately 3 weeks. Tissue held in a shortened range for longer than 3 weeks has undergone contractural changes that must be addressed before attempts at shift correction.

Finally, this technique may be used to decompress compressive lesions such as nerve impingement syndromes. Aside from backward bending, side bending is the least stressful movement on the disc, followed in increasing order of stress by forward bending and, finally, rotation. In rehabilitation of discogenic lesions, the side bending elongation maneuver decompresses the nerve root and takes the disc into the next most stressful maneuver.

Patient position: Sidelying.

Therapist position: The therapist stands perpendicular over the patient. The top forearm contacts the lateral thorax/rib cage, while the bottom forearm contacts the lateral portion of the ilium.

Hands: The fingers contact the medial border of the erector spinae.

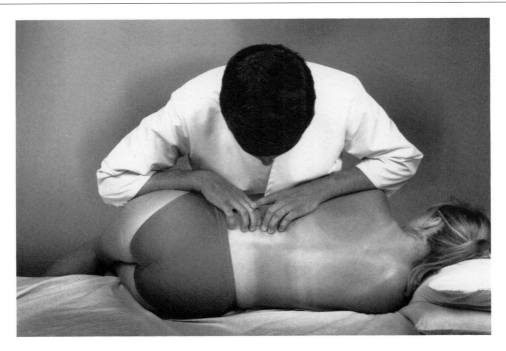

Figure 8–28

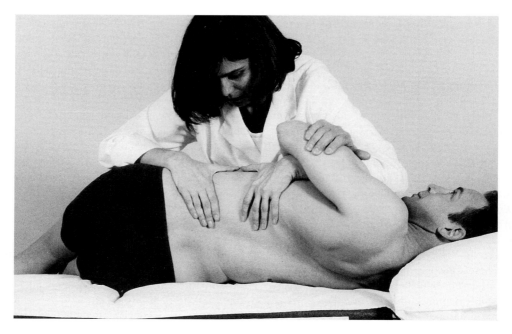

Figure 8–29

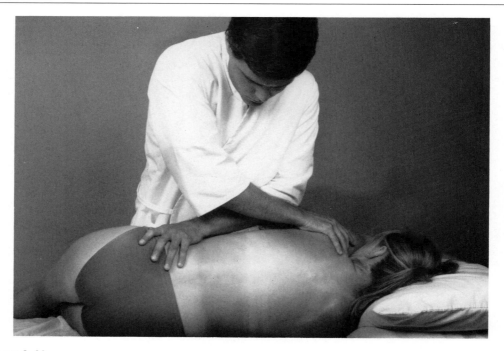

Figure 8–30

Execution: To localize forces in the lumbar area, the hips and knees are bent to 90 degrees and the patient's feet are allowed to hang off the table. Care must be taken while lowering the feet off the table not to provoke any symptoms. Once the feet are off the table, pressure is exerted in a cephalic direction with the top forearm and in a caudal direction with the bottom forearm. At the same time, the fingers move from medial to lateral on the erector spinae. The forearms are localizing most of the stretch on the quadratus. The hands are primarily aiding this movement by gently releasing the erector spinae.

In this position, a gentle hold-relax technique may be performed by asking the patient to gently push the ilium into the therapist's bottom forearm. The patient should not be allowed to remain with the legs off the table for more than 30 to 45 seconds, since the lever arms of the lower extremity are applying considerable forces into the lumbar spine.

To diffuse the forces and provide a more general elongation of the lumbar and thoracic spines, the patient is asked to fully flex the shoulder and hold the top of the treatment table. The legs are then lowered off the table, as described. The forces may be applied through the arm-hand contacts described above, or a traction-elongation force may be applied through the palms of the hands as shown in Figure 8–30. The therapist can apply an elongation of the lateral connective tissue of the lumbar and thoracic spines and even into the connective tissues of the shoulder girdle complex.

In some cases, where the quadratus lumborum has been hypertonic, but not necessarily short-ened, it may be necessary to create more length in the quadratus than the previously described technique. In order to manufacture more length, the trunk is rotated to the T12/L1 segment. By rotating in this fashion, rib 12 is rotated away from the pelvis, allowing for lengthening of the more cephalic aspect of the quadratus lumbo-rum. Once rotated, the legs are placed off the table and a sidebending force is placed on the pelvis as previously described. The top arm continues to sidebend at approximately 30 degrees

off center. The therapist should not continue to rotate. The change in angle of the sidebending provides a more aggressive stretch of the quadratus lumborum. **Note of caution:** Discogenic lesions are a strong precaution here, because the rotation could compromise a discogenic lesion.

Forward Bending Laminar Release (Figures 8–31 and 8–32)

Purpose: The purpose of this technique is to elongate the posterior myofascial tissues of the lumbar spine. This may be necessary in hyperlor-dotic postures or in preparation for joint mobili-zations. As discussed earlier, soft tissue and joint mobilization have a unique relationship in that either the soft tissues or the joint may be contrib-uting to a hypomobility. Passive segmental mo-bility of a joint may change dramatically after releasing soft tissue. On the other hand, joint mobilization may have a profound effect on the surrounding myofascial tissues by way of stimu-lating joint receptors. This technique is often performed before, during, and after joint mobili-zation to complement specific joint maneuvers.

Patient position: The patient is sidelying in a semifetal position.

Therapist position: The therapist stands per-pendicular over the patient. The therapist will stabilize the patient's top knee by placing it in the area of the therapist's anterior hip for control and ease of execution.

Hands: The top hand is placed over the tho-racolumbar junction, along with the forearm in such a way that the elbow is positioned in a cephalic direction while the fingers are posi-tioned in a caudal position. The top hand is the "stabilizing hand." The bottom hand is placed initially in the area of the upper lumbar spine in contact with the erector spinae, with the fingers slightly flexed.

Execution: To execute the technique, the fin-gers are moved caudally down the length of the erector spinae while the patient's hip is simul-taneously being flexed. The leg movement is executed through the therapist's hip and pelvis. The therapist pulls the patient's knee toward

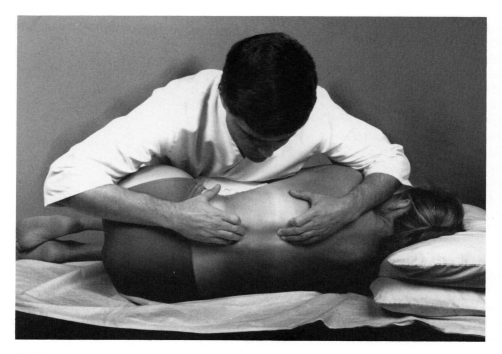

Figure 8–31

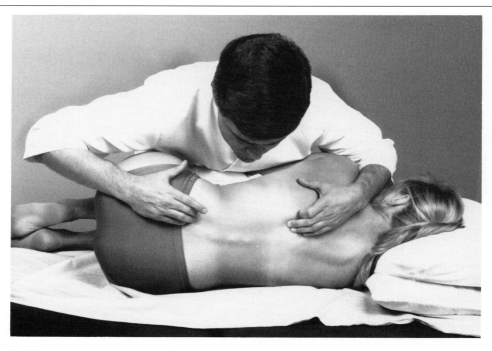

Figure 8–32

the chest, decreasing the lumbar lordosis. This allows for elongation coming from hip flexion as well as from the caudal stroking of the therapist's bottom hand.

If a specific joint restriction is found, this technique may be somewhat localized to prepare the surrounding soft tissues prior to a joint mobilization. The hip is first flexed to the level of the restriction. During this time the fingers are palpating between the spinous processes for the forward bending restriction. Once movement has arrived at the appropriate level, the hip is extended slightly to reslacken the tissue at that level. The top stabilizing hand is brought down to a position just cephalic to the restricted level. The bottom hand is brought up to a level almost contacting the top hand. The therapist then strokes over the erector spinae in a caudal direction the length of 2 to 3 segments while the hip is being flexed through a short arc of movement. This allows for tissue to be elongated both by the hip flexion and by the caudal pull of the bottom hand. Passive intervertebral mobility should be assessed prior to an appropriate number of repetitions of this technique.

Longitudinal Posterior Hip Release

Purpose: This technique is an extension of the previous technique, but is sometimes used separately for lesions in the area of the posterior hip. Piriformal lesions and parasacral lesions, as well as extensibility problems in the posterior hip, are effectively treated with this technique.

Patient position: The patient is sidelying in the semifetal position with the top knee stabilized in the anterior hip of the therapist.

Therapist position: The therapist is standing perpendicular to the patient, stabilizing the patient's top knee with the anterior hip, allowing for an effective mechanical advantage for the therapist, and a feeling of security for the patient.

Hands: The top hand is placed so that the palm gently contacts the ASIS. The bottom hand is positioned over the buttock with the fingertips just distal to the SIJ.

Execution: The top hand gently stabilizes the ilium while the patient's hip is gently flexed. The therapist accomplishes this by leaning in a cephalic direction with his or her pelvis. Simultaneously, the bottom hand strokes from just distal to the PSIS to the ischial tuberosity and laterally in a paratrochanteric direction. Most of the pressure is applied through the fingertips, but the palm remains in contact throughout.

Forward Bending Laminar Release—All Fours (Figures 8–33, 8–34, and 8–35)

Purpose: The purpose of this technique is to elongate the posterior soft tissues of the lumbar or thoracic spines. This technique may serve as an alternative to the forward bending laminar release in sidelying. If the patient is too large for the therapist to manage in sidelying, the all-fours position may be used. Specificity is sacrificed somewhat in order to gain some mechanical advantage. One advantage to this technique is that the patient actively participates rather than remaining passive.

Patient position: Quadruped.

Therapist position: The therapist stands at the patient's side at a 45-degree angle. The therapist may need to be on a stool, or, if a high-low table is used, the table should be lowered.

Hands: For optimal stability and efficiency, the thumb is held against the PIP joint of the index finger. Contact is made using both the PIP joint and the tip of the thumb (Figure 8–2).

Execution: The therapist instructs the patient to bend forward first at the cervical spine and gradually recruit motion into the thoracic spine. As movement is recruited into the thoracic spine, the therapist asks the patient to start rocking back on his or her heels. This motion begins to recruit movement from lower lumbar to upper lumbar areas. As the patient recruits this movement, the therapist longitudinally strokes the erector spinae unilaterally with the bottom hand, starting from the sacrum and moving toward the thoracolumbar junction. The top hand is used as a guiding hand to dictate the quantity and pace of the patient's movement.

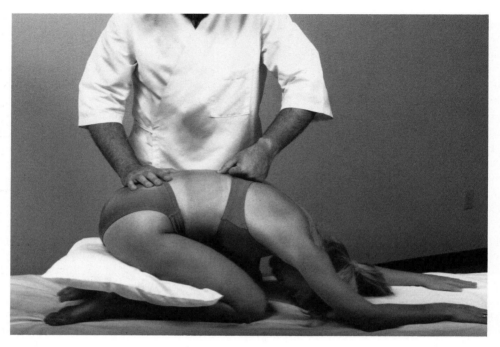

Figure 8–33

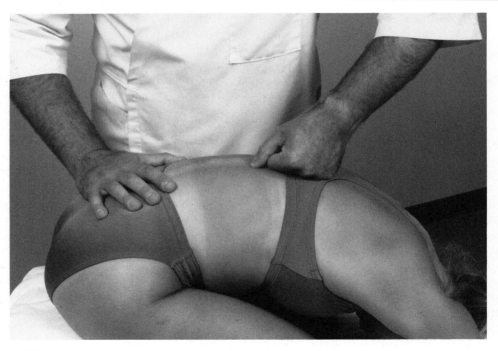

Figure 8–34

Figure 8–35

Forward Bending Laminar Release—Sitting (Figures 8–36, 8–37, 8–38, and 8–39)

Purpose: This technique will elongate the posterior myofascial structures of the lumbar, thoracic, and to a certain extent, cervical spines. As with the quadruped technique, the patient actively participates in the technique; the technique also allows for working with patients larger than the therapist. Specificity is somewhat sacrificed, but significant mechanical advantage is gained in performing the technique in a sitting position.

Patient position: Sitting.

Therapist position: The therapist stands behind the patient, facing the patient.

Hands: The hand placement is as illustrated in Figure 8–2. The position with the thumb held next to the PIP joint of the index finger is a very stable position and does not compromise the joints of the hand.

Execution: The patient is first asked to forward bend segmentally starting from the cervical spine, recruiting into the thoracic spine, and finally into the lumbar spine. Once the patient understands the concept of segmental movement, the thumb-PIP complex of each hand is placed over the erector spinae at the cervicothoracic junction in a downward position. For optimal mechanical advantage, the elbows should be directed upward, and the thumb-PIP should be directed downward. The patient is asked to forward bend segmentally, and the therapist strokes the erector spinae longitudinally at the level the movement is being recruited. If a localized restriction is found, the patient may be asked to stop the movement at the point of the restriction, and a sustained pressure may be applied.

The same technique may be applied unilaterally and with a rotatory component by asking the patient to forward bend diagonally. The patient

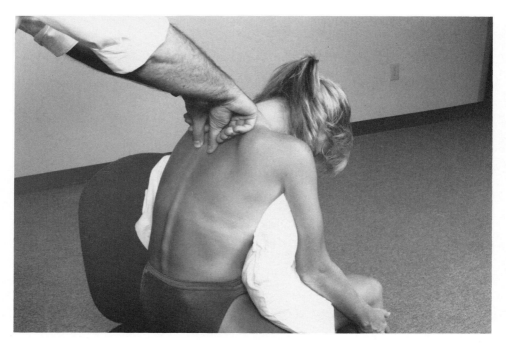

Figure 8–36

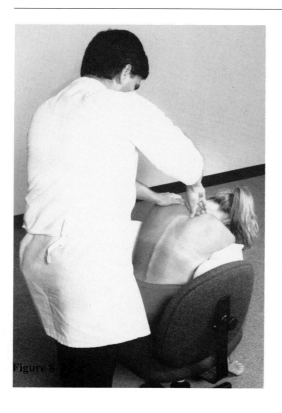

is asked to follow the lateral border of the leg with the arms. This maneuver allows for forward bending, side bending, and rotation components. One hand is used as a guidance hand to dictate the pace and quantity of movement, and the other hand is used to perform the technique. The patient is again asked to move segmentally into the diagonal plane, and the therapist strokes the erector spinae at the level that movement is being recruited. If a movement restriction and/or myofascial restriction is encountered, the patient may be asked to stop, and the therapist may apply a sustained pressure.

The technique may also be applied to the cervical spine. The therapist uses one hand to guide the patient's head and neck, generally into a diagonal direction, and uses the other hand to stroke down the cervical paravertebral muscles.

Contraindications: This technique should not be used with discogenic backs since a loaded spine is being taken into forward bending.

Figure 8-

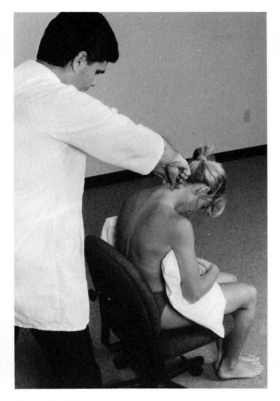

Figure 8–38

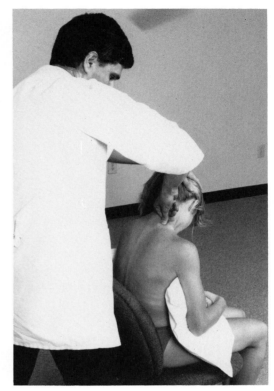

Figure 8–39

Lumbar Myofascial Roll (Figure 8–40)

Purpose: This technique is an excellent preparatory technique for a midlumbar roll mobilization or manipulation. Many times, a midlumbar joint manipulation is difficult to execute because of myofascial restrictions or active muscle guarding. The patient may be apprehensive of rotating the spine to the degree that is required in the midlumbar manipulation. Decreasing myofascial restrictions not only allows the patient to relax into rotation, but also facilitates locking a specific joint of the lumbar spine.

Patient position: Sidelying.

Therapist position: The therapist stands facing the patient at the level of the lumbar spine. The top hand is placed over the patient's subclavicular-pectoral area, while the bottom hand is placed over the midlumbar area. The knee of the patient is placed in the anterior portion of the therapist's hip.

Hands: The fingers of the bottom hand are placed on the medial aspect of the erector spinae.

Execution: The lumbar spine is bent forward by flexing the patient's hip and recruiting motion into the lumbar spine. The lumbar spine is then rotated by pulling the bottom arm of the patient until movement is recruited into the lumbar spine. In the therapist position described above, the lumbar spine is rotated from both contact points. The erector spinae muscles are simulta-

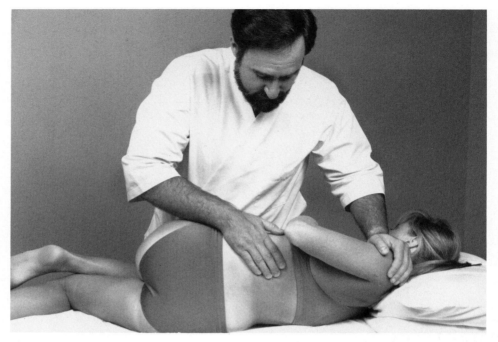

Figure 8–40

neously stroked diagonally with the fingers as the rotatory force is applied through the top arm. The lumbar spine may be rotated close to end range, but should not be taken to the limit of motion. As relaxation and elongation are achieved, the spine may be taken to end range to perform the joint manipulation.

Lateral Shear (Figures 8–41 and 8–42)

Purpose: This technique is performed to normalize the lateral shear forces in the lumbar spine, which may be abnormal and/or asymmetrical due to past trauma. An excellent use of this technique is with a resolving discogenic lesion where the patient has ceased experiencing a lateral shift for a period of time. When the patient is tested for lateral shear (passively shifted), the patient will usually adopt the position of the previous shift quite easily, and will be markedly restricted when sheared in the opposite direction. Normalizing this myofascial imbalance is the primary purpose of the technique.

This technique should not be confused with the lateral shift correction technique, which is typically performed on a laterally shifted patient. The technique has application for neuromuscular retraining at end-stage discogenic rehabilitation, but should not be used early in the discogenic rehabilitation process, especially when a lateral shift is still present. The technique of choice in a lateral shift is the lateral shift correction technique.

Test procedure: To determine if a lateral shear imbalance exists, the therapist stands behind the patient and passively moves the patient into a lateral shift position. This is accomplished by placing one hand on the ilium and the other hand on the upper trapezius-shoulder girdle area and applying force. The force on the ilium is directly lateral (i.e., in a horizontal plane), while the pressure applied on the upper

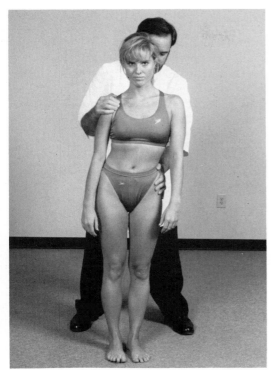

Figure 8–41

trapezius-shoulder girdle is in a 45-degree diagonal direction. The vector on the upper trapezius/shoulder girdle is a combination of lateral force (in the horizontal plane) and compressive force. If the patient's trunk moves easily to the right and is restricted in movement to the left, the patient is restricted in left lateral shear.

The next step is determining whether the restriction is merely postural or if a true myofascial restriction exists. The patient then lies prone and the lateral shear is again tested, this time primarily from the pelvis. If the patient's pelvis moves easily to the left and is restricted in movement to the right, the patient is said to be restricted in left lateral shear. Remember, the direction of the shear is always based on the direction the upper body moves in relation to the lower body. In standing, if the trunk is restricted in movement to the left, a left lateral shear restriction exists. In the prone position, ilial movement to the right is trunk motion to the left. If ilial movement to the right is restricted, the restriction is still in left lateral shift.

If a movement restriction exists when the patient stands but normalizes when the patient is

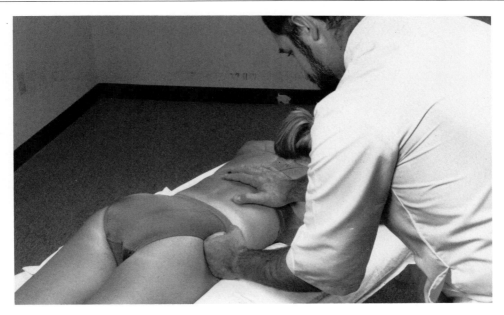

Figure 8–42

prone, the condition is not as significant, and is usually more easily treated. If a movement restriction exists when standing and remains when prone, the condition has become more entrenched and can potentially be more detrimental if left unchecked. Either way, treatment is necessary to correct the dysfunction.

Patient position: Prone.

Therapist position: The therapist stands perpendicular to and over the patient at pelvis level.

Hands: The palm of the hand or a fist may be used to make contact on the ilium, just proximal to the greater trochanter of the hip.

Execution: The restriction is engaged by gently shearing the pelvis laterally. Once resistance is met, the patient is asked to hold his or her position, and then relax (hold-relax stretch). As the patient relaxes, the pelvis is sheared farther laterally and the process is repeated. After several repetitions, the lateral shear is retested,

both in prone and in standing positions, to see if the technique produced any change.

Two things are accomplished in this technique. The first is a neuromuscular "repassing" to eliminate muscular holding patterns created by old trauma. The second is releasing restrictions in the noncontractile elements that became restricted as a result of prolonged dysfunction in the contractile elements.

Diaphragm (Figures 8–43, 8–44, 8–45, and 8–46)

Purpose: These techniques are designed to free up restrictions in the anterior fascia just caudal to the rib cage, and to mobilize the diaphragm. In a forward-head, protracted shoulder, slumped position, the anterior elements collapse, reducing diaphragmatic excursion. This can lead to increased activity in the secondary accessory breathing muscles. Also, for the pa-

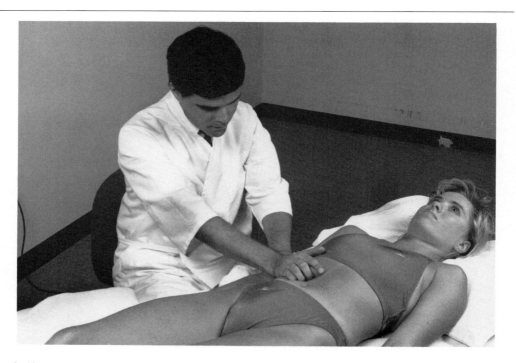

Figure 8–43

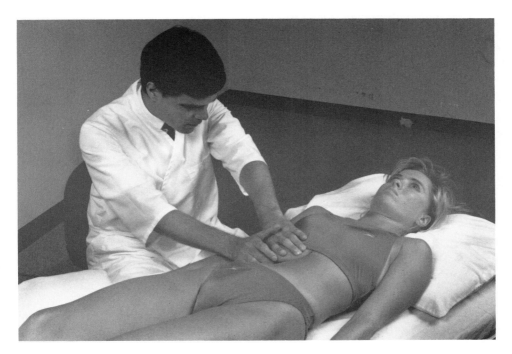

Figure 8–44

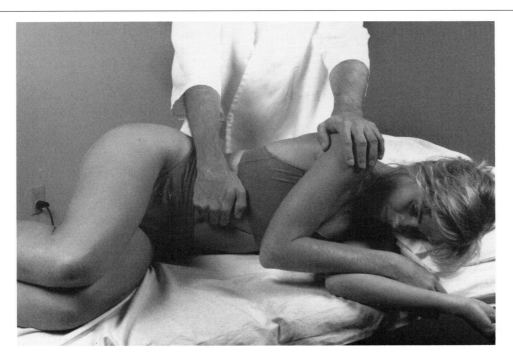

Figure 8–45

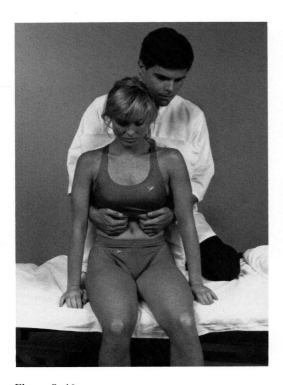

Figure 8–46

tient to perform postural reeducation techniques successfully and elongate the thoracic area, the contracted area of the anterior chest and abdomen must be supple and mobile. Three techniques are shown, ranging from the least aggressive to the most aggressive; the general progression should follow the patient's tolerance level.

First Position: Supine

Patient position: The patient lies in the supine position with the knees and hips slightly flexed.

Therapist position: The therapist is either standing at the side of the patient or seated. The seated position is biomechanically more advantageous for the therapist.

Hands: The therapist's top hand is placed over the bottom portion of the rib cage. The bottom hand is placed at the anterior-medial border of the rib cage, just lateral to the xiphoid process, and in the connective tissue just caudal to the rib cage.

Execution: The top hand gently pushes the connective tissue in a caudal direction in order to slacken the tissue just caudal to the rib cage. This allows the fingers of the bottom hand to slide underneath the rib cage (to patient tolerance). The stroke is applied, following the border of the rib cage medial to lateral. Care should be taken not to push into the floating ribs while moving laterally with the stroke. In this position, only a superficial or moderate level of penetration can be achieved.

Second Position: Sidelying

Patient position: The patient is in the sidelying position with the hips and knees flexed to 90 degrees.

Therapist position: The therapist stands behind the patient.

Hands: The hand position is similar to that described above. The top hand is placed on the lower portion of the rib cage, while the bottom hand is placed at the caudal border of the rib cage, just lateral to the xiphoid process.

Execution: With the patient more flexed, more slack is placed in the superficial connective tissue. The first technique actually mobilizes both the connective tissue and the diaphragm. The second technique bypasses the superficial connective tissue to engage the deeper connective tissue under the rib cage. The therapist uses the top hand once again to move the connective tissue medially and caudally, allowing the bottom hand to slide under the rib cage. The stroke is again applied in a medial to lateral direction, with care not to hit the floating ribs.

Third Position: Sitting

Patient position: The beginning position for this technique is the slumped sitting posture. This allows the therapist greater access to the tissues underneath the rib cage. As the technique is performed, however, the patient may be asked to assume a more erect posture so the therapist can mobilize the rib cage.

Therapist position: The therapist stands behind the patient with a pillow between the therapist and the patient. The patient is leaning into the therapist in a slumped posture.

Hands: Whereas the previous techniques are unilateral, this technique is bilateral. Both hands slide underneath the rib cage medially, just lateral to the xiphoid process.

Execution: The stroke is again executed medial to lateral with the patient in the slumped position. At an appropriate time, the hands firmly grip the rib cage, and the patient is asked to inhale deeply and attempt a more erect posture. The rib cage is mobilized anteriorly.

Psoas (Figures 8–47, 8–48, and 8–49)

Purpose: Mobilization of the psoas muscle is clearly indicated in cases where actual shortening exists, which is creating mobility problems in the lumbar spine, especially with forward bending. In an axially extended posture (flat back posture), however, the psoas may be hypertonic in an effort to increase lordosis or to guard a lesion, where axial flexion of the lumbar spine is the primary dysfunction producing symptoms.

Patient position: The patient lies in the supine position with the hips and knees flexed approximately 30 to 45 degrees. This puts the muscle in a slackened position. If the muscle does not exhibit enough slack, the hips may be flexed 90 degrees, over the therapist's leg. This should be performed on a high-low table for optimal biomechanical advantage.

Therapist position: The therapist stands at the patient's side, and if necessary, places one leg on the table; the patient's legs are then placed over the therapist's leg. The therapist may use the leg to change the amount of hip flexion during application of the technique.

Hands: The fingertips are used to contact the psoas. The hands are placed lateral to the umbilicus and the psoas is approached from a 45-degree angle.

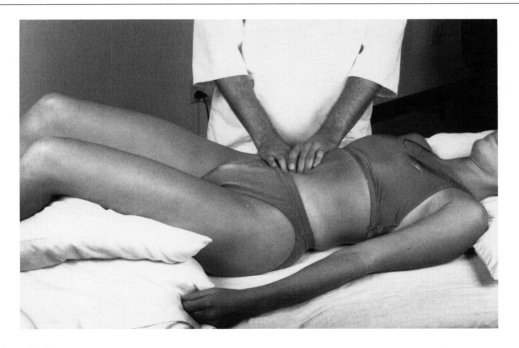

Figure 8–47

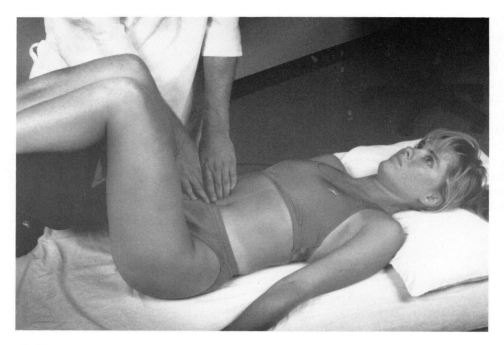

Figure 8–48

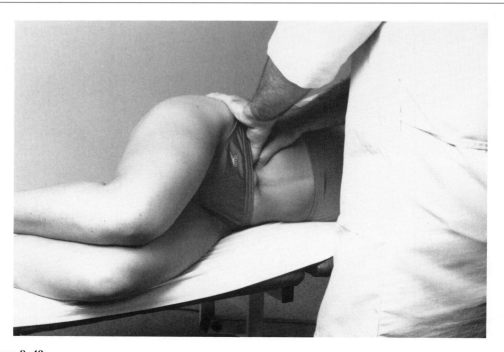

Figure 8–49

Execution: Because of the location of the psoas, a significant depth must be achieved through the abdomen. Care must be taken to progress slowly into the appropriate depth, asking the patient about the relative comfort of the technique. As more depth is achieved through the abdomen, "landing" on a more rigid structure indicates arrival onto the psoas. The psoas will be more rigid than the soft tissue of the abdomen. The patient will also report a different sensation, usually more noxious when the psoas is palpated, especially if the psoas is dysfunctional.

Because longitudinal stroking of a muscle is generally less noxious than cross stroking, the psoas should be gently stroked longitudinally at first. Only after longitudinal stroking should a cross stroking of the psoas be attempted. Once the technique is terminated, the hands should be gradually removed from the abdomen. In some cases, the psoas may be more accessible with the patient in a sidelying position. The therapist may use the thumbs to access and release the psoas.

Iliacus (Figures 8–50 and 8–51)

Purpose: The iliacus muscle can be treated for limited extension of the hip or as an extension of a psoas release. Even though the iliacus does not have an insertion into the spine, a shortening dysfunction of the iliacus can anteriorly rotate the pelvis, creating a backward bending dysfunction of the spine.

Patient position: The patient lies in the supine position with the hip flexed approximately 30 degrees. As with the psoas, if not enough slack is placed on the tissue, the hip may be flexed by the therapist, up to approximately 110 degrees.

Therapist position: Standing over the patient, and if necessary grasping the lower extremity to impart hip flexion.

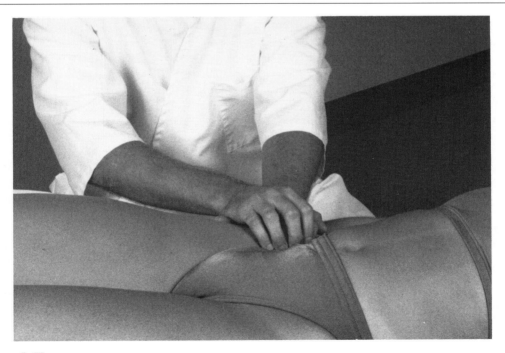

Figure 8–50

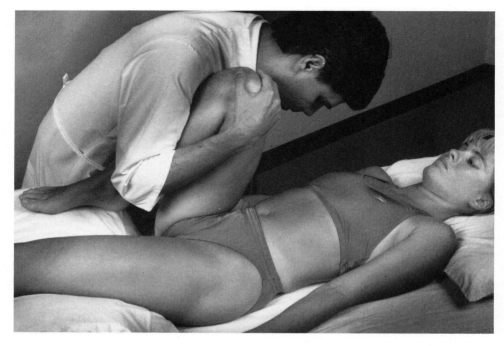

Figure 8–51

Hands: The palm of the hand is placed over the anterior superior iliac spine and the fingers are wrapped over the ilium, contacting the anterior surface of the ilium. The fingers are in contact with the iliacus at the most accessible portion of the insertion.

Execution: The technique begins with a proximal to distal stroking of the muscle (longitudinal stroking). As patient tolerance or muscle response dictates, the stroke is shifted into a cross stroking of the iliacus (lateral to medial).

TECHNIQUES FOR THE LUMBOPELVIC/LOWER QUARTER AREA

Greater Trochanter Rocking (Figures 8–52A,B and 8–53)

Purpose: This technique is designed for gentle inhibition of the lateral rotators of the hip as well as for the hamstrings. This is an excellent preparatory technique for more exten-sive work in the piriformis, posterior hip, and hamstrings. Application of this technique will generally yield an increase in straight leg raising as well as internal rotation.

Patient position: The patient is in the supine position.

Therapist position: The therapist is either standing or seated at the patient's side.

Hands: The fingers of the top hand will contact the posterior surface of the greater trochanter, while the bottom hand gently grasps the leg in the area of the distal femur, just proximal to the knee joint.

Execution: A gentle internal rotation motion is begun with the bottom hand. Simultaneously, an anterior pressure is applied with the top hand through the greater trochanter, further facilitating the internal rotation motion. The motion is repeated in an oscillatory fashion at a deliberate speed. The technique is generally performed in the midrange of internal rotation and is gradually moved toward end range. Internal rotation

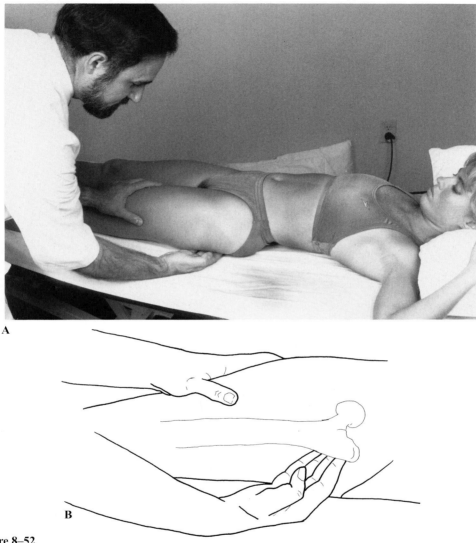

Figure 8–52

and straight leg raising should be reassessed after this technique.

Transverse Muscle Play of Quadriceps (Figures 8–54, 8–55, and 8–56)

Purpose: The concept of muscle play is applied to the quadriceps muscle where the surrounding fasciae are mobilized to provide more room for the quadriceps to contract. The "bending of the water hose" analogy applies in this case. The technique has a different "look" compared to the muscle play of the erector spinae because of the size of the quadriceps compared to that of the erector spinae.

Patient position: Supine or sidelying.

Therapist position: The therapist stands at the patient's side at the level of the midfemur.

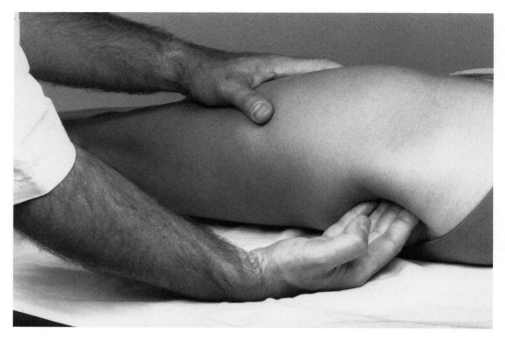

Figure 8–53

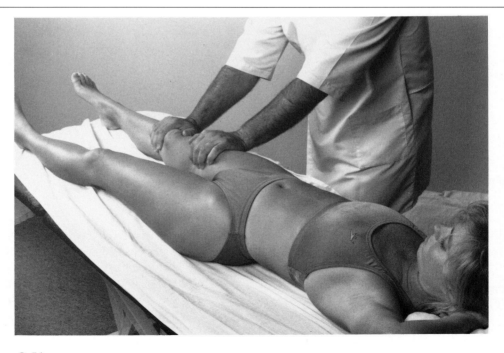

Figure 8–54

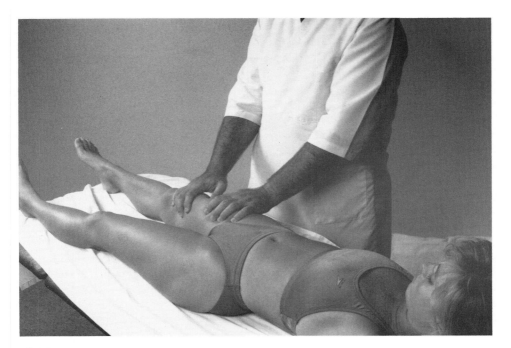

Figure 8–55

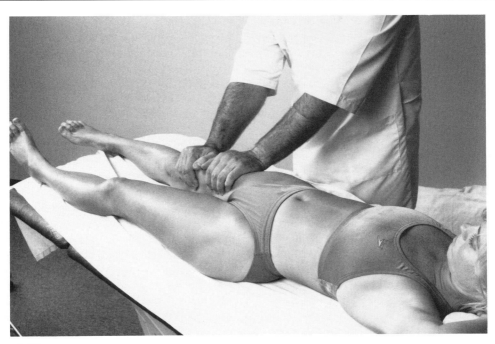

Figure 8–56

Hands: The bottom hand grasps the quadriceps and femur distally, just proximal to the knee joint. The top hand grasps the quadriceps anywhere on the muscle belly where a restriction is identified. The top hand palm is placed laterally over the vastus lateralis. Alternately, both hands may be placed over the quadriceps to engage more surface area.

Execution: Firmly grasping the distal aspect of the quadriceps with the bottom hand, the top hand shears the quadriceps from lateral to medial over the femur. The force is applied through the palm of the hand. The hand does not slide over the skin, however. The technique is designed to move the muscle, not to slide over the muscle, which is more of a massage technique. The technique is generally performed in a lateral to medial direction since more restrictions seem to occur in the vastus lateralis. The technique may be performed in a medial to lateral direction by moving to the patient's other side and proceeding to shear the quadriceps in a medial to lateral direction. The technique may also be performed in diagonal planes if a restriction occurs in that plane.

The main difference between soft tissue mobilization and joint mobilization is that in joint mobilization, arthrokinematic rules must be followed. In soft tissue mobilization, restrictions may occur in any plane and at any depth, and mobilization of the restriction does not depend on arthrokinematics.

Iliotibial Band Paratrochanteric Mobilization (Figures 8–57A,B; 8–58A,B; 8–59, and 8–60)

Purpose: The iliotibial band is an area commonly involved in lower kinetic chain problems, knee dysfunction, and hip and low-back dysfunction. Many diffuse "referred pain" syndromes in the lower extremity can be traced to iliotibial dysfunctions. Treatment of this area becomes important to a variety of problems, even if the patient has no conscious awareness of pain in the area. Many times the patient will be exquisitely tender over the area of the iliotibial band and surrounding tissues when other dysfunctions are symptomatic nearby.

This technique actually addresses three distinct areas: (1) the connective tissue "groove" between the iliotibial band and the hamstring, (2) the groove between the iliotibial band and the quadriceps, and (3) the iliotibial band itself. Because loose irregular connective tissue is the most easily mobilized, the surrounding connective tissue will more readily respond than the iliotibial band.

The other area this technique addresses is the paratrochanteric area. The connective tissue surrounding the greater trochanter is also often dysfunctional; this includes superior, inferior, anterior, and posterior to the greater trochanter.

Patient position: (1) Patient lies supine with the hip and knee flexed, but with the foot on the treatment table. (2) In a more aggressive form of the technique, the patient is asked to flex and adduct the hip and to hold the position to place the posterior hip in a more stretched position. The execution of the technique is the same in either position.

Therapist position: The therapist stands at the patient's side at a slight angle to the patient, depending on whether the anterior or posterior border of the iliotibial band is being treated.

Hands: The hand position described previously in Figure 8–2 is used in this technique. The thumb and the PIP of the index finger contact one another and become the point of contact with the patient. The elbow should point up toward the ceiling for the best mechanical advantage in applying the technique.

Execution: (1) Posterior border of iliotibial band. The therapist's top hand stabilizes the patient's leg at the knee joint. The thumb and PIP of the bottom hand contact the groove between the iliotibial band (ITB) and the hamstring distally. With the elbow pointing upward, the stroke follows the border of the ITB and the hamstring proximally. When the area of the greater trochanter is reached, the direction of the stroke changes and continues paratrochanterically to encircle the greater trochanter. (2) Anterior border of

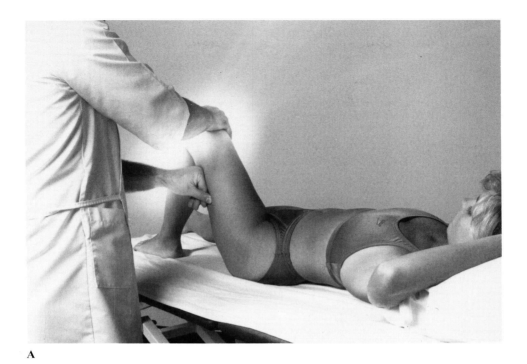

A

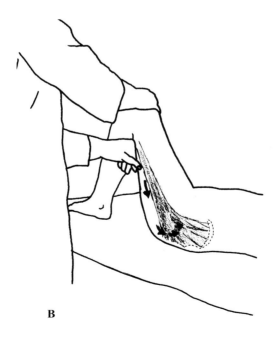

B

Figure 8–57

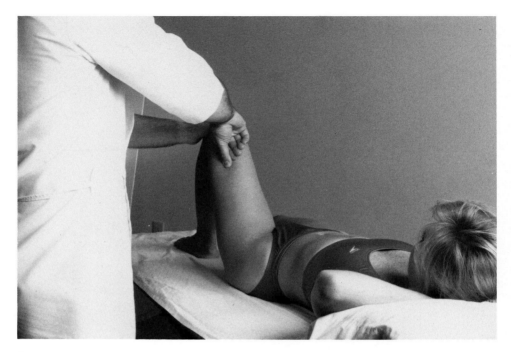

A

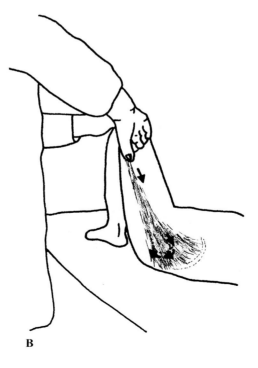

B

Figure 8–58

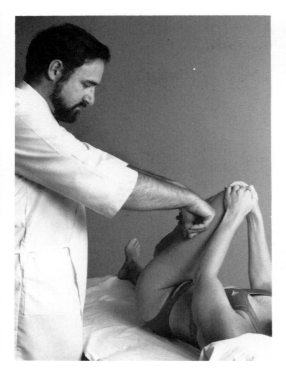

Figure 8–59

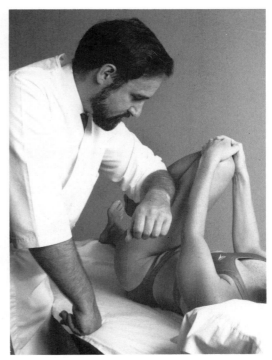

Figure 8–60

the iliotibial band. The therapist's bottom hand stabilizes the patient's leg at the knee joint. The thumb and PIP of the top hand contact the groove between the iliotibial band and the quadriceps distally. With the elbow pointing upward, the stroke follows the border of the ITB proximally, again until the greater trochanter is reached. The stroke continues over the anterior border of the greater trochanter, encircling the greater trochanter and ending posteriorly. (3) Direct technique over the iliotibial band. The therapist's bottom hand stabilizes the patient's leg at the knee. The elbow contacts the ITB and the stroke proceeds from distal to proximal directly over the ITB and greater trochanter. Both hands may also be used to stroke directly over the ITB. The above techniques may be repeated with the patient holding the leg in hip flexion

and adduction to stretch the posterior elements of the hip, and for greater access to the ITB proximally.

Hold-Relax Stretch of Hip (Figure 8–61)

Purpose: The purpose of this technique is to stretch the posterior hip capsule and surrounding periarticular soft tissues. A typical patient presentation is a middle-aged man with a flattened lumbar lordosis, hypermobile lumbar facet joints, tight hamstrings, and restricted posterior hip connective tissues. With little pelvic contribution to forward bending, the lumbar spine becomes progressively more hypermobile and symptomatic. Facet as well as disc degeneration may result as a long-term effect. The focus of treatment lies in establishing a balance between

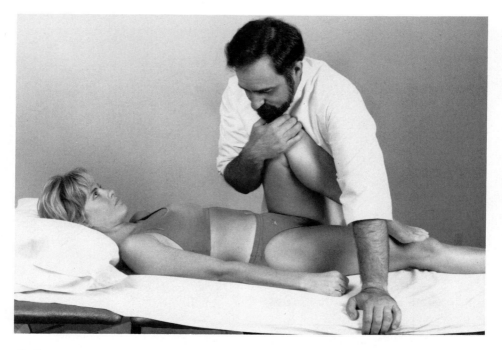

Figure 8–61

the low back and the hip in forward bending. To accomplish greater movement balance, the periarticular structures of the hip must be mobilized before movement reeducation can begin.

The technique of choice to prepare the tissue for this procedure is the paratrochanteric technique described above. Paratrochanteric mobilization will prepare the tissue for aggressive stretching.

Patient position: The patient is in the supine position with the hip flexed and adducted.

Therapist position: The therapist stands over the patient, facing the patient. The patient's leg is placed so it is in contact with the therapist's chest. The knee should approximate the therapist's axillary or pectoral area.

Hands: Both hands are grasping the treatment table on either side of the table, "strapping" the patient to the table, or one hand can grasp the patient's leg for added stability.

Execution: With the patient in the therapist's firm grasp, the patient is asked to push the leg into the therapist's chest. The patient is then asked to release the contraction and the therapist "takes up the slack," moving the hip into further flexion-adduction. Occasionally, the patient will complain of anterior hip pain while the technique is being executed. A possible explanation is that the anterior capsule may be pinching with the extreme amount of flexion being applied to the hip. An alternate execution of the technique is to bring the hip out of extreme flexion and to emphasize the technique's adduction component. The therapist stabilizes the pelvis at the ASIS with the top hand. The leg is grasped with the bottom arm, and adducted with a slight external rotation component. The addition of external rotation and the increase in adduction will compensate for the loss of flexion and regain the tissue tension lost with the loss of hip flexion.

Hamstrings (Figures 8–62, 8–63A,B; 8–64, and 8–65A,B)

Purpose: The purpose of these techniques is to mobilize the hamstrings in preparation for aggressive stretching technique. The hamstrings may be restricted in a longitudinal direction, medial lateral direction, or in a diagonal plane. By identifying and treating lesions in the appropriate plane and position, specific restrictions may be released and flexibility of the hamstrings may be increased prior to stretching.

Patient position: Supine with the hip and knee flexed approximately 90 degrees, and resting over the shoulder of the therapist.

Therapist position: The therapist is seated on the treatment table facing the patient.

Longitudinal Stroking

Hands: Contact with the patient is made with the "fist" (i.e., with the MCP joints of the hand), or with the elbow. Contact is first made on the distal aspect of the hamstrings.

Execution: With the patient's leg relaxed over the therapist's shoulder, firm pressure is applied with the fist or elbow to the distal aspect of the hamstrings. The hamstrings are stroked longitudinally, distal to proximal to the insertion at the ischial tuberosity. If the restriction lies in the proximal hamstring near the ischial tuberosity, the hip may be flexed beyond 90 degrees.

If a specific restriction is identified, the elbow may be used to apply a sustained pressure on the restriction. The stroke should be stopped when the restricted area is reached. The pressure should be sustained for an appropriate period until changes in the restriction are palpable, or until it is obvious that no change is going to occur.

Splay Technique

Hands: The hands gently grasp the middle aspect of the lower extremity so the thumbs are

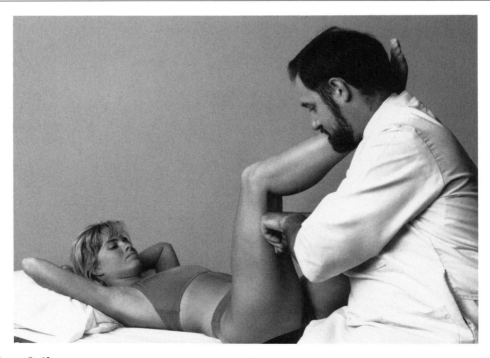

Figure 8–62

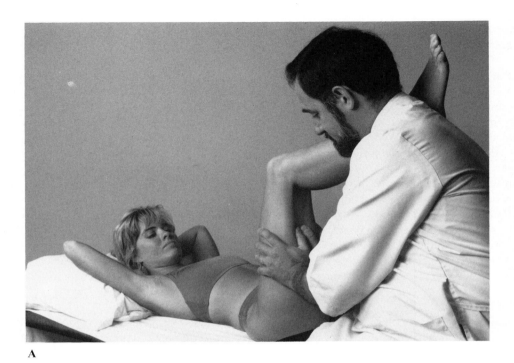

A

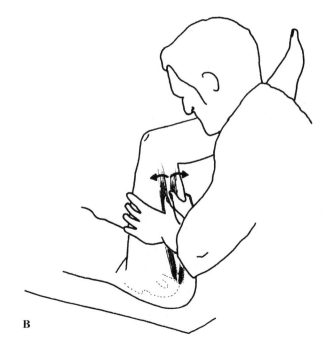

B

Figure 8–63

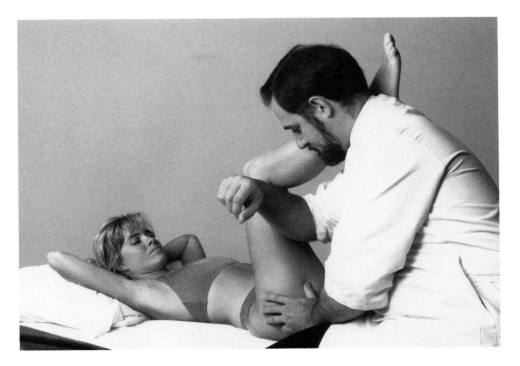

Figure 8–64

in contact with the distal portion of the hamstrings. The thumbs approximate one another at the medial aspect of the lower extremity.

Execution: Deep pressure is applied medially by the thumbs, as the hamstrings are stroked longitudinally from proximal to distal. As the distal portion of the hamstrings is reached, the stroke direction changes to medial/lateral, splaying or pulling the hamstrings apart. The thumbs do not slide over the hamstring muscle bellies. Rather, the thumbs are grasping the muscle bellies and pulling them apart. This technique can be thought of as a specific form of muscle play for the distal hamstrings.

Stretch of Proximal Hamstring (Figure 8–66)

Purpose: The purpose of this technique is to isolate a stretch of the proximal hamstring. Hamstring injuries generally fall into two basic categories: mid belly injuries and proximal injuries. Proximal injuries can be more serious, more recurring, and more difficult to treat than mid belly lesions. The proximal injury can sometimes act similar to an "epicondylitis," where the injury is in the tenoperiostial junction. By isolating a stretch to the proximal hamstring, the therapist can more effectively aid in the remodeling of the proximal tissues.

Patient position: Supine, with the leg resting on the therapist's shoulder.

Therapist position: The therapist stands on one leg and places the other leg on the treatment table. The patient's leg is placed comfortably on the therapist's shoulder.

Hands: The therapist places his/her hands around the knee of the patient. This will help to provide a traction force and control the amount of knee flexion.

Execution: The therapist first performs a straight leg raise until the patient feels a mild hamstring stretch. The patient is then asked to

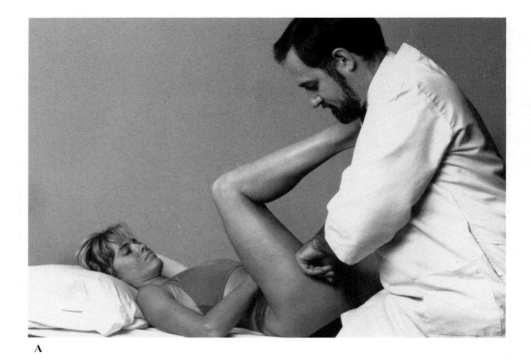

A

B

Figure 8–65

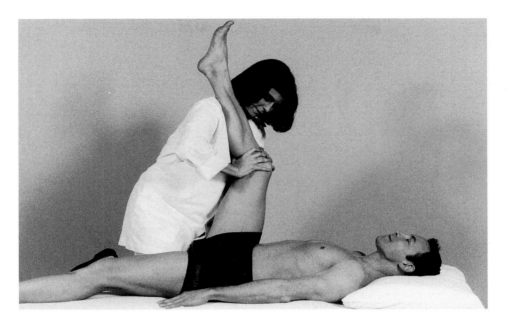

Figure 8–66

localize the stretch. If the stretch is felt in the distal or mid belly of the hamstring, the therapist allows the patient's knee to bend slightly. Keeping the slight bend constant, the therapist continues to flex the hip until the patient again feels the stretch. At this point, the patient should feel the stretch more proximally because the distal aspect has been slackened and the proximal aspect has been further stretched. The therapist repeats the process, allowing the knee to flex slightly more, and then flexing the hip further. The process is repeated until the stretch is felt closest to the origin at the ischial tuberosity. To further localize the stretch, a slight traction force can be placed on the leg while stretching. The traction serves to pull slightly more on the origin of the muscle at the ischial tuberosity.

Cross-Friction Ischial Tuberosity—Greater Trochanter (Figure 8–67)

Purpose: Many hamstring injuries and/or dysfunctions occur at the junctional zone (i.e.,

the insertion of the hamstrings into the ischial tuberosity). Healing and restoration of proper function may be facilitated with a deep cross-frictional type of mobilization over this area.

Patient position: Prone.

Therapist position: Standing over the patient in a diagonal position.

Hands: The fingertips or the tips of the thumbs may be used for this technique. The most stable position of the hands for application of the technique is the four-finger position previously described in the iliac crest technique. The fingers are placed over the insertion of the hamstrings, just distal to the ischial tuberosity.

Execution: The fingers palpate deeply until firm pressure is placed on the hamstring insertion and junctional zone. The fingers are oscillated medial to lateral consistent with the concept of cross-friction. The fingers are then moved proximally onto the ischial tuberosity. The periosteum of the ischial tuberosity may be also damaged or dysfunctional. The same medial to lateral movement is applied over the ischial

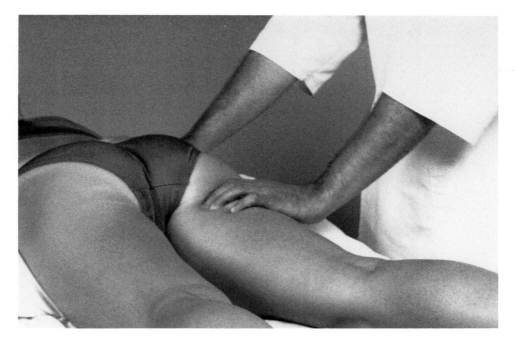

Figure 8–67

tuberosity. This technique should be applied aggressively to the point where it is seminoxious to the patient.

Fascial Plane between Ischial Tuberosity and Greater Trochanter (Figures 8–68 and 8–69)

Purpose: A fascial plane or connective tissue sheath exists in the area between the ischial tuberosity that, when restricted, may limit hip extension. Since the greater trochanter moves anteriorly with hip extension, restrictions in this fascial sheath may limit hip extension. The area is not usually painful and rarely tender, but may create hip or lumbar dysfunctions if not extensible.

Patient position: The patient lies in the prone position. The hip may be held or positioned in the extended position in order to add tension to the tissue.

Therapist position: The therapist stands over the patient in a diagonal position. If the therapist chooses to extend the hip manually (as opposed to positioning the hip with pillows), the lower extremity is grasped with the bottom hand, leaving the top hand free to execute the technique. If the lower extremity is not held by the therapist, both hands should be used in executing the technique.

Hands: The hand position described in the iliac crest release technique is used. Both index and ring fingers approximated together provide the stability necessary to perform a technique at this depth. The pressure is exerted through the fingertips.

Execution: The direction of force is primarily in a posterior to anterior direction, with a slight horizontal component. As in the iliac crest release technique, an oscillatory motion is performed repetitively in an anterior direction. In order to apply tension to the fascial sheath, the

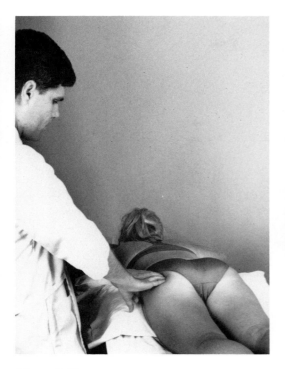

Figure 8–68

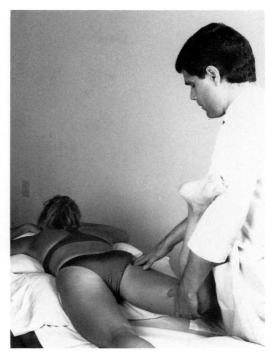

Figure 8–69

hip may be extended by the therapist or positioned on pillows. Following the technique, the hip may be stretched into extension as a follow-up technique.

Friction of Piriformis Insertion (Figure 8–70)

Purpose: This technique helps prepare the piriformis for direct contact on the muscle belly if the piriformis muscle is reactive and cannot tolerate direct pressure, or if direct pressure is not resulting in any palpable changes or changes in symptoms.

Patient position: Prone.

Therapist position: The therapist stands at the patient's side at the level of the hip.

Hands: The bottom hand grasps the leg at the ankle and bends the knee to 90 degrees. The thumb of the top hand is placed on the superior border of the greater trochanter. The superior border of the greater trochanter is palpated by gently internally and externally rotating the leg with the bottom hand. The thumb is placed in the soft tissue above the lateral aspect of the greater trochanter. As the hip is gently internally and externally rotated, the thumb moves distally until arrival at the first bony prominence. The prominence is the superior border of the greater trochanter.

Execution: Once in position, the thumb does not move. The technique is applied by midrange and pain-free rotation of the hip. As the rotation occurs, the thumb will come on and off the greater trochanter. A fairly deep pressure is applied, but only to patient tolerance. Care must be taken not to take the hip into excessive internal rotation if the piriformis is very reactive.

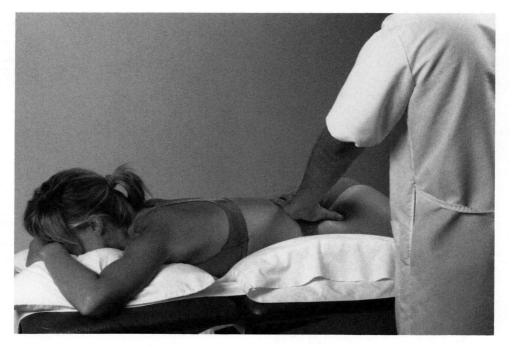

Figure 8–70

Piriformis Release in Prone (Figures 8–71, 8–72, and 8–73)

Purpose: This technique is used in cases where the dysfunction lies in a hypertonic muscular state of the piriformis rather than in a connective tissue dysfunctional state. The technique is primarily designed to decrease underlying muscle tone, and secondarily to affect connective tissue. The technique is performed in a graded fashion depending on the overall pain and reactivity of the piriformis muscle.

The issue is raised here whether the "piriformis syndrome" exists or not. Some say that the syndrome does not exist, but the average clinician, in practice, cannot deny the involvement of the piriformis or manifestations of piriformis hypertonicity. The clinical reality is that "piriformis syndrome" in a pure sense is rare, but piriformis involvement related to other dysfunctions is seen quite often.

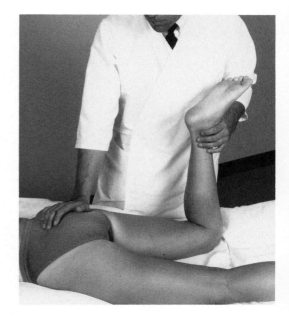

Figure 8–71

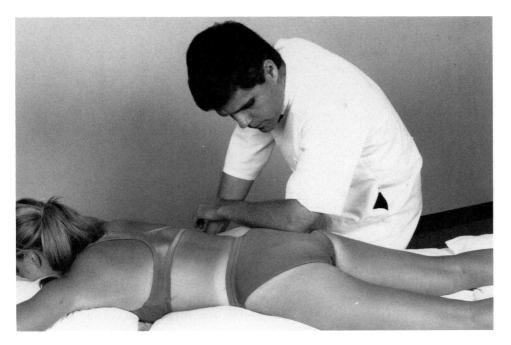

Figure 8–72

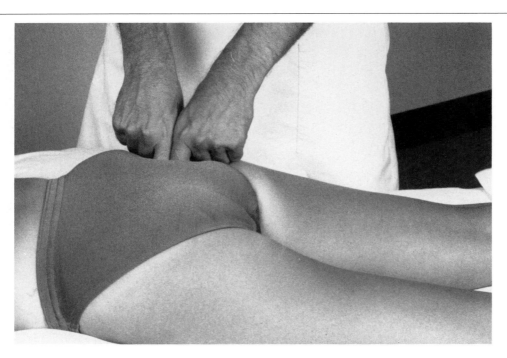

Figure 8–73

Patient position: Prone.

Therapist position: The therapist stands at the patient's side, perpendicular to the patient.

Hands: The hand position in the technique will vary depending on the reactivity of the muscle and the tolerance of the patient. The general progression of the technique goes through three different hand positions: (1) palm of the hand, (2) elbow, and (3) PIP joints of both hands.

Execution: (1) Using the palm of the hand, the therapist applies gentle pressure at mid buttock, which is the general location of the mid belly of the piriformis. The leg is gently externally rotated to put the piriformis on slack. The pressure is gently increased until the level of the piriformis is reached. A sustained pressure is applied, provided the pressure does not create an increase in tone. As the piriformis relaxes, more pressure can be progressively applied. If the piriformis releases, even partially, the patient tolerance will increase, allowing the next variation of the technique. (2) The same sustained pressure may be applied to the piriformis using the elbow. The elbow allows for more localized pressure to be applied. The same principle applies in that as the piriformis releases and as the pain decreases, more pressure can be applied. (3) Finally, the PIP joints of both hands may be used to apply even more localized pressure. If the patient is able to tolerate it, a gentle oscillatory motion can be performed to inhibit further and mechanically mobilize the piriformis.

Transverse Muscle Play of Hamstrings (Figures 8–74 and 8–75)

Purpose: As described for the quadriceps, the concept of transverse muscle play can be used to mobilize the fascial sheath surrounding the hamstrings to provide more space for the hamstrings to contract.

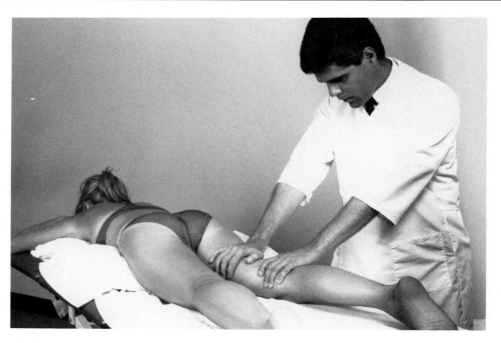

Figure 8–74

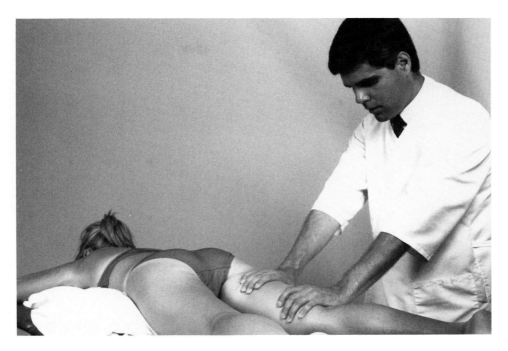

Figure 8–75

Patient position: Prone.

Therapist position: The therapist stands at the patient's side at the level of the mid femur.

Hands: The bottom hand grasps the hamstrings and femur distally, just proximal to the knee joint. The top hand grasps the hamstrings anywhere on the muscle belly where a restriction is identified. The palm of the hand is initially placed over the lateral hamstring, just posterior to the ITB. Both hands may also be used to gain a greater contact surface.

Execution: Grasping the distal aspect of the hamstrings with the bottom hand, the top hand shears the hamstrings in a lateral to medial direction, with major force being applied through the palm of the hand. The hand does not slide over the skin. The technique may also be performed in a medial to lateral direction if the restriction is present in that direction. The therapist should approach the patient from the other side of the table so a medial to lateral force may be applied

with the palm of the top hand. If a restriction is felt in a posterior-anterior direction in the medial hamstring, the force may be applied in a posterior to anterior direction, again with the palm of the hand. Remember, restrictions can occur in any direction or plane, and the technique direction should be modified to treat the restriction adequately.

Transverse Muscle Play of Adductor Muscles (Figure 8–76)

Purpose: As previously described in concept, the technique is designed to mobilize the surrounding fascial sheaths of the adductor muscles. This is an excellent preparatory technique for adductor stretching.

Patient position: Prone.

Therapist position: The therapist stands at the patient's side, holding the leg with the knee bent at 90 degrees.

Figure 8–76

Hands: The palm of the top hand is used to apply the transverse pressure on the adductor group.

Execution: The palm of the hand makes contact with the adductor muscles and partially with the medial hamstring. Pressure is applied toward the treatment table to create the bending movement of the adductors.

Transverse Muscle Play of Gastrocnemius-soleus (Figures 8–77 and 8–78)

Purpose: The fascial sheath surrounding the gastrocnemius-soleus muscle group is mobilized in order to increase extensibility and allow for more efficient contraction of the muscle group. Longitudinal stretching is also facilitated after application of this technique.

Patient position: Prone.

Therapist position: The therapist stands at the level of mid tibia.

Hands: The bottom hand grasps the distal aspect of the gastrocnemius-soleus muscle group just proximal to the Achilles tendon. The top hand grasps the gastrocnemius-soleus muscle group at the level of the muscle where the restriction is identified. As before, both hands may be used to attain a more optimal "bend" in the muscle.

Execution: Grasping the distal aspect of the gastrocnemius-soleus muscle group firmly with the bottom hand, the top hand shears the muscle from lateral to medial with the palm of the hand. The hand does not slide over the skin. Sliding over the skin modifies the technique into a pure massage technique. As with the other techniques, the technique may be performed medial to lateral, or posterior to anterior, depending

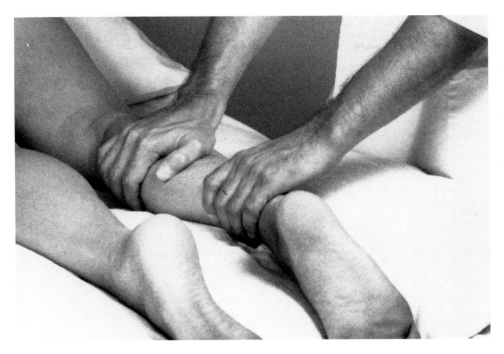

Figure 8–77

Figure 8–78

on the direction of the restriction. The clinician should be sensitive to restrictions and follow them with the technique, since no arthrokinematic rules apply. The success of the treatment often will depend on whether or not the direction of application was properly identified.

Bony Clearing of the Tibia (Figure 8–79)

Purpose: The purpose of this technique is to clear fascia from the anterior and posterior compartments as they adhere to the tibia. Many lower kinetic chain problems, especially in athletes participating in ballistic sports (running, basketball, soccer etc.), develop fascial adhesions related to "shin splints." The bony clearing techniques are effective in mobilizing the fascia as it adheres to the tibia. This technique can be used for both anterior and posterior compartmental syndromes.

Patient position: Supine.

Therapist position: Standing or sitting at the foot of the table.

Hands: The thumb pushes off the border of the tibia, creating a "wedge" between the bone and the approximating soft tissue. The thumb is positioned either anterior or posterior, depending on the compartment that is affected.

Execution: A small amount of lubrication is used. The thumb drives a wedge between the bone and the approximating soft tissues distally. The thumb then moves proximally, continuing to stay in the wedge, and also continuing to approximate the tibia. In compromised areas, the wedge will either not be as deep, or have adhesions that make the wedge nonexistent. These are adhesions that need to be mobilized.

For the posterior side, the knee may be bent, and the foot placed on the table to allow for slightly more slack in the tissues.

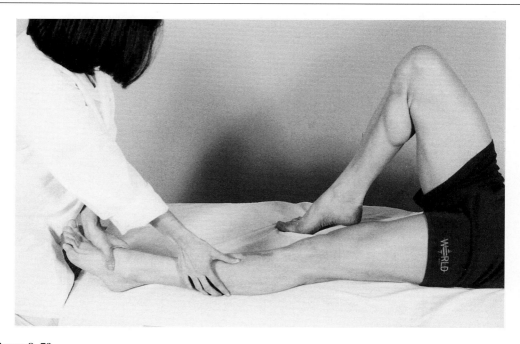

Figure 8–79

Lateral Fascial Distraction of the Tibia (Figure 8–80)

Purpose: The purpose of this technique is to stretch the posterior compartment fascia that is adhered to the tibia laterally. As with the technique above, this technique will be effective in the treatment of lower leg compartment syndromes, shin splints, etc. that are caused by excessive ballistic lower kinetic chain activity.

Patient position: Prone with the knee flexed to 90 degrees and plantarflexed slightly.

Therapist position: Seated on the side of the table at the patient's lower leg.

Hands: The lateral hand is placed distally and will be used as a counter lever. The palm of the medial hand is placed on the mid belly of the gastrocnemius-soleus muscle group as close to the tibia as possible without actually contacting it.

Execution: The therapist puts a medial to lateral pressure on the gastrocnemius-soleus muscle group, pulling it away from the tibia. The technique starts in the mid belly, but can move proximal or distal, depending on the location and severity of the restriction. The therapist carefully attempts to push the muscle laterally into the plastic range, keeping an eye on patient reaction. This technique can be quite painful if the fascia along the tibial/gastrocnemius border is compromised.

Cross Friction of the Gastrocnemius-soleus Musculotendinous Junction (Figure 8–81)

Purpose: Many patients involved in ballistic type sport activities develop fascial thickening in the musculotendinous junction of the gastrocnemius-soleus muscle group. This phenomenon

Figure 8–80

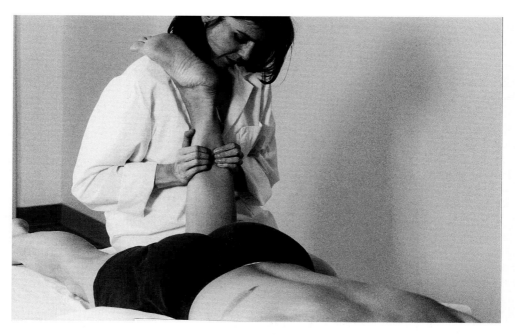

Figure 8–81

may occur with or without muscular shortening. The purpose of this technique is to mobilize the musculotendinous junction and the fascia immediately surrounding it.

Patient position: Prone with the knee flexed to 90 degrees and the foot plantarflexed moderately.

Therapist position: Seated at the side of the table at the lower leg of the patient.

Hands: The hands gently grasp the lower leg so that the fingers come to rest directly over the musculotendinous junction of the gastrocnemius-soleus muscle group.

Execution: The therapist applies firm pressure over the musculotendinous junction with the fingers and applies a firm cross frictional movement across the junction, watching for patient response. This area can be exquisitely tender in active patients participating in ballistic type sporting activities. Note that the tissue is held in the shortened range. Again, this is to create slack and allow for access to deeper tis-

sues. A stretch can immediately follow the application of this technique.

TECHNIQUES FOR THE THORACIC/UPPER THORACIC SPINE AND UPPER EXTREMITY

Lateral Elongation of Upper Thoracic Area (Figures 8–82, 8–83, 8–84, and 8–85)

Purpose: The purpose of this technique is elongation of the soft tissue structures of the upper thoracic area (posterior and anterior). The technique is especially applicable for patients with protracted shoulder girdle complexes and forward-head postures. After application of the technique, the shoulder girdle and upper thoracic spine assume a more relaxed and retracted position. This technique should be used before attempting postural reeducation techniques. Initially, the clinician emphasizes both the anterior and posterior structures of the upper thoracic

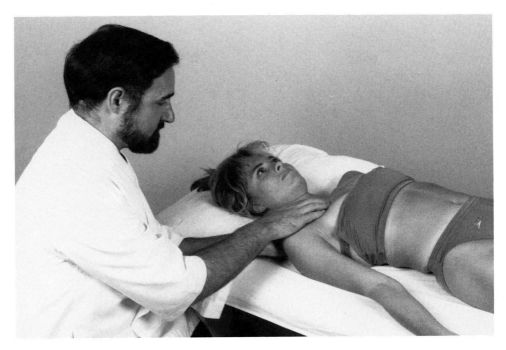

Figure 8–82

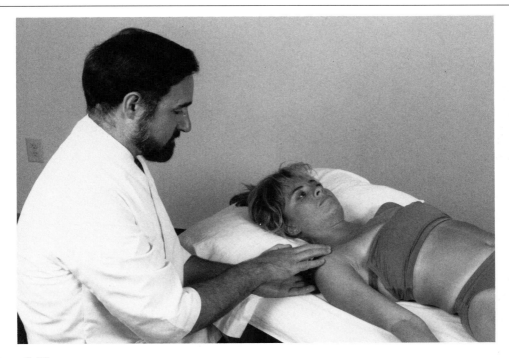

Figure 8–83

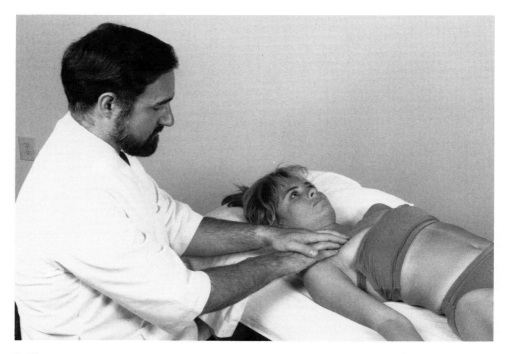

Figure 8–84

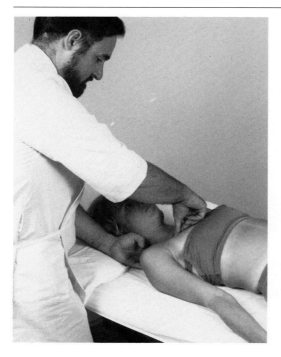

Figure 8–85

area. As the technique progresses, more emphasis is placed on the anterior structures. Three alternate hand placements are described, each of which progresses into deeper tissues of the anterior chest.

Patient position: The patient is supine with the head lying flat on the treatment table.

Therapist position: The therapist is seated at the head of the table, at a 45-degree angle to the patient.

Anterior-Posterior Technique

Hands: One hand is placed posteriorly, so that the fingertips are just lateral to the spinous processes of the upper thoracic spine. The hand should be resting superior to the spine of the scapula. The other hand is placed infraclavicularly, with the fingertips just lateral to the sternum.

Execution: The primary force of the technique comes from the fingertips, even though contact is maintained through the palm of the hands. The stroke begins medially and progresses laterally, as the therapist pulls the hands toward the glenohumeral joint. Once the stroke is completed, the hands are quickly placed in the start position again and the stroke is repeated. The pressure is placed through each hand and is moderate in depth.

Deep Anterior Technique

Hands: To approximate deeper structures, both hands are placed anteriorly. One hand is placed over the other, again over the infraclavicular area. The fingertips are just lateral to the sternum.

Execution: The stroke is applied through the fingertips from medial to lateral. Deeper pressure is applied through the hands and fingertips.

Rib Splaying: Ribs 1–3 (This aspect of the technique is the most aggressive

form. The depth of penetration is to the intercostal spaces.)

Therapist position: Standing, facing the patient.

Hands: Contact will be made with the thumb and PIP of the index finger as shown in Figure 8–2.

Execution: The stroke begins medially in the intercostal space of the 1st and 2nd ribs. The intercostal space is followed laterally until no longer palpable (a short distance). The stroke is performed in intercostal space of ribs 2 and 3 (in men in the intercostal space of ribs 3 and 4).

Unilateral Posterior/Anterior Articulation of First Rib (Figure 8–86)

Purpose: This technique is technically a joint mobilization technique, but blends in well with the above techniques, especially if rib dysfunc-

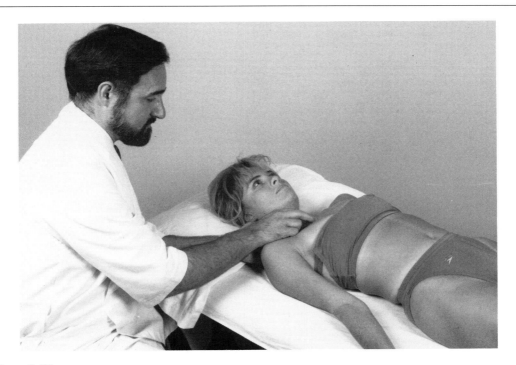

Figure 8–86

tion is present. With increased myofascial tone in the subclavicular area, the upper thoracic area, and the scalenes, joint mechanics in the first rib can easily become dysfunctional. The purpose of this technique is not to change the position of the first rib, but to increase mobility.

Patient position: The patient lies supine with the head flat on the treatment table.

Therapist position: The therapist is seated at the head of the table, at a 45-degree angle to the patient.

Hands: The bottom hand (which is usually the hand closest to the patient) palpates the posterior aspect of the first rib near the costotransverse junction. This can be accomplished by first palpating the posterior aspect of the upper trapezius. The clinician then continues caudally and medially until bone is palpated. This bone is the first rib. If the finger is too lateral, the border of the scapula is palpated; if the finger is too medial, the spinous process will be palpated. The top hand palpates just lateral to the first sternocostal articulation. The clinician may first palpate the sternoclavicular junction with the middle finger and slip the finger just caudal and lateral, which is just lateral to the first sternocostal junction.

Execution: The clinician applies a moderate oscillatory movement anterior/posterior and posterior/anterior. Enough pressure should be applied to create movement in the first rib. The rate of oscillation should be 2 to 3 oscillations per second.

First Rib Shoulder Depression Technique (Figure 8–87)

Purpose: This technique is largely inhibitory in nature, although the first rib is being gently articulated. The rhythm created by the rib and

Figure 8–87

shoulder articulation provides a form of bio-feedback for the patient, and can indicate to the clinician and patient the degree of inherent relaxation or tension in the upper thoracic area. This subtle form of biofeedback releases tone in the upper thoracic area, preparing the tissue for deeper or more specific myofascial work, and facilitates joint mobilization and manipulation.

Patient position: The patient lies supine with the head flat on the treatment table.

Therapist position: The therapist is seated at the head of the table at a 45-degree angle to the patient.

Hands: The hand closest to the patient palpates the posterior aspect of the first rib as described in the previous technique. Palpating the posterior aspect of the upper trapezius, the clinician then continues caudally and medially until bone is palpated. This bone is the first rib. If the finger is too lateral, the border of the scapula is palpated, and if the finger is too medial, the spinous process will be palpated. The other hand is placed on the superior aspect of the shoulder joint complex.

Execution: Execution of this technique involves two separate movements occurring simultaneously: (1) With the bottom hand, the rib is articulated anteriorly; (2) with the other hand, the shoulder is depressed caudally. The two motions occur simultaneously in a slow deliberate rhythm (approximately 2 oscillations per second). During execution, the patient may become aware of increased tone, tension, or holding patterns, and may spontaneously relax. The tissue is then prepared for other techniques as necessary.

Bilateral Upper Thoracic Release (Figure 8–88)

Purpose: The purpose of this technique is to release the deep paravertebral musculature

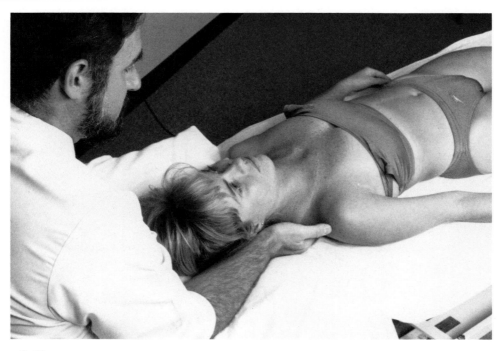

Figure 8–88

of the upper thoracic spine. The technique is accomplished in two distinct maneuvers. The first is a moderate depth, cephalic-caudal movement, and the second is a deep anterior/posterior movement.

Patient position: The patient is supine with the head flat on the table.

Therapist position: The therapist is seated at the head of the table directly behind the patient.

Hands: The hands slide onto the paravertebral musculature of the upper thoracic spine (to approximately T4). The fingers make firm contact with the paravertebral musculature.

Execution: The first maneuver is a gentle cephalic-caudal oscillation with moderately deep pressure on the upper thoracic paravertebrals. The oscillations should be performed at a rate of approximately 2 per second. In the second maneuver, the direction of the movement changes from cephalic-caudal to anterior articulations. While this may be considered anterior/posterior mobilization of the upper thoracic spine, the

firm pressure applied through the layers of muscle onto the deep muscle provides adequate force to release deep underlying tone. The clinician should exercise caution in guarding his or her hands, since this technique requires maximum force through the fingers. Fatigue will occur quickly and the clinician should proceed to another technique. Efficiency and ease of application of technique are essential for effective technique delivery. Any strain or inefficiency on the clinician's part will be transferred to the patient, and reduce the potential effect of the technique.

Pectoralis Major Muscle Play—Pectoralis Minor (Figures 8–89, 8–90, and 8–91)

Pectoralis Major

Purpose: In the forward-head posture, the pectoralis major and minor become restricted and shortened. This creates an inability to stand or sit erect without significant effort from the

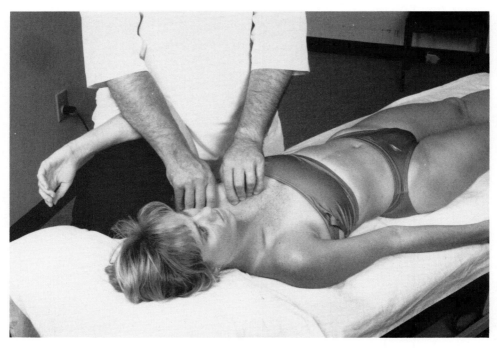

Figure 8–89

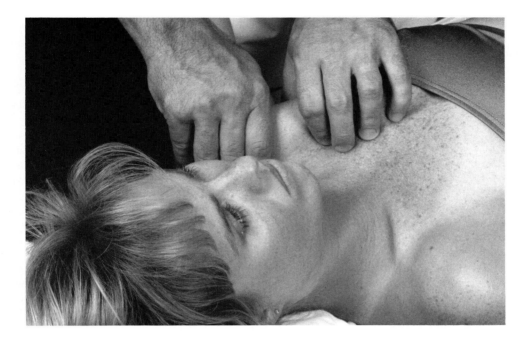

Figure 8–90

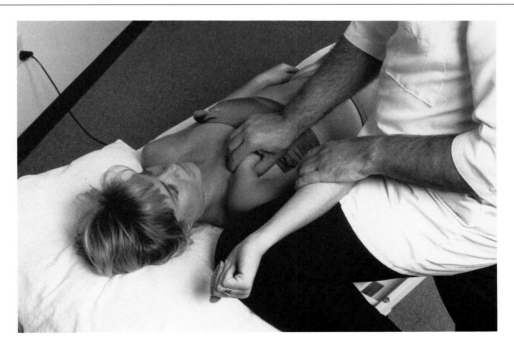

Figure 8–91

patient. Before postural reeducation can occur effectively, the pectorals must have adequate extensibility.

Patient position: The patient is in the supine position. The shoulder is flexed 90 to 120 degrees.

Therapist position: The therapist is standing over the patient at a 45-degree angle to the patient. The therapist may place a leg on the table to allow the patient's arm to rest in a relaxed position.

Hands: The thumbs slide underneath the pectoralis major, and the hands grasp the muscle firmly between the thumbs and fingers.

Execution: The technique can be likened to the garden hose analogy in which a garden hose is being bent. The pectoralis muscle is grasped firmly between the thumbs and fingers and is gently lifted or bent away from the thorax. The movement can be a sustained movement or an oscillatory movement.

Pectoralis Minor

Hands: With one hand maintaining the same position as described above, the thumbs are moved posteriorly until in contact with the pectoralis minor. The muscle may be difficult to palpate, but if the ribs are palpable, the muscle is being palpated.

Execution: The thumbs are pressed onto the pectoralis minor, and a gentle "cross-friction type" technique may be performed. Care must be taken because the pectoralis minor area is very tender even if not dysfunctional.

Seated Pectoral Anterior Fascial Stretch (Figures 8–92 and 8–93)

Purpose: The purpose of this technique is to stretch the anterior structures (fascia, pectoralis major, minor) to allow for more erect posture.

Patient position: Seated, with hands behind head, or with elbows straight.

Figure 8–92

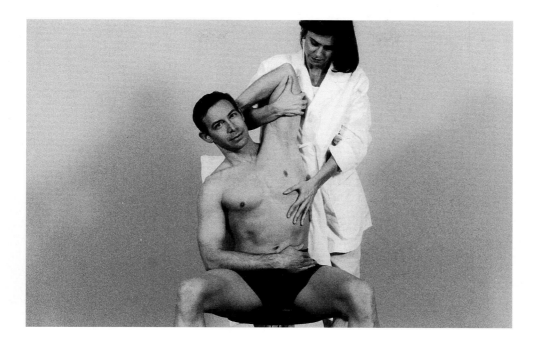

Figure 8–93

Therapist position: The therapist is standing behind the patient with either his or her hip or knee stabilizing the thoracic spine and acting as a fulcrum. A pillow should be placed between the patient and the therapist.

Hands: Bilateral Stretch: The hands will grasp the middle part of the upper arm. Unilateral Stretch: The inside hand of the therapist grasps the upper part of the patient's arm. The outside hand is place on the middle part of the antero-lateral rib cage.

Execution: Bilateral Stretch: The pressure is applied in a lateral, posterior, and cephalic direction for maximum elongation. The patient is asked to breathe deeply to increase elongation anteriorly.

Unilateral Stretch: Using the inside arm and body, the patient's arm is pulled posteriorly and superiorly, stretching the anterior fascia. The

outside hand on the rib cage is pushed caudally to further engage the anterior superficial fascia.

Subscapularis (Figures 8–94 and 8–95)

Purpose: The subscapularis is generally not an area reported by the patient to be painful. The area may be significantly restricted and extremely tender to palpation, however. Since the internal rotators are held in a shortened position in the forward-head protracted shoulder posture, the subscapularis and the surrounding myofascia become restricted, acting as barriers to efficient postural reeducation.

Patient position: The patient is in the supine position with the shoulder flexed from 90 to 170 degrees, depending on the restriction and comfort level of the patient.

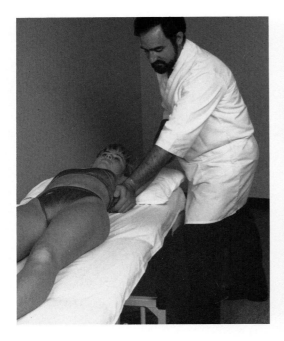

Figure 8–94

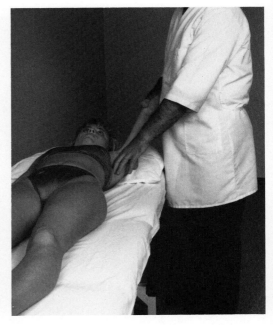

Figure 8–95

Therapist position: The therapist is standing at the head of the table at a 45-degree angle to the patient. The patient's arm is grasped by the therapist close to the therapist's body to provide a slight traction-distraction force.

Hands: The hands may be placed on the patient in three different ways, depending on how aggressively the therapist wishes to deliver the technique. The palm of the hand, the thumb, or the fingertips may be used in order from least aggressive to most aggressive.

Execution: (1) The patient's arm, which is in some degree of flexion, is gently distracted. The palm of the other hand is placed on the lateral border of the scapula, as close to the glenohumeral joint as possible. As gentle distraction is placed on the arm, the palm strokes caudally and toward the inferior angle of the scapula.

If fascial restrictions exist, the stroke may be lengthened to include the lateral fascial sheaths between the scapula and the ilium.

(2) In the same position, the thumb is used to stroke caudally. Thumb placement is more specific, being located on the anterior surface of the lateral border of the scapula. The arm is again distracted and the thumb moves caudally over the anterolateral border of the scapula toward the inferior angle.

(3) Finally, specific restrictions, either in the lateral aspect of the subscapularis or in the fascial sheath between the scapula and the thorax, may be treated using the fingertips. The tips of the index, middle, and ring fingers palpate the anterior surface of the lateral scapula and gentle pressure is applied. The pressure may be sustained or slow oscillatory in nature.

Anterolateral Fascial Elongation (Figures 8–96 and 8–97)

Purpose: The anterior fascial planes are often restricted, especially in the slumped posture or in various shoulder pathologies. The purpose of this technique is to elongate the superficial fascial sheaths of the anterior thorax.

Patient position: The patient is in the supine position, with the shoulder flexed 120 to 170 degrees.

Therapist position: The therapist stands behind the patient, grasping the patient's arm and providing a distraction of the arm.

Hands: The entire surface of the hand is placed just below the nipple line. (Note: Male therapists treating female patients should carefully drape the patient and should stay well below breast tissue.)

Execution: As the arm is tractioned into flexion, a traction force is applied to the superficial fascia, first in the direction of the umbilicus.

The direction of the force may be changed, and directed more diagonally toward the contralateral ASIS or into a more cardinal plane direction toward the ipsilateral ASIS. The shoulder should be in as much flexion as possible to allow for maximal stretch of the connective tissues. The use of skin lubricants for this technique is discouraged.

Anterolateral Fascial Elongation with Rotational Component

Purpose: If the myofascia is restricted in a rotational direction, the above technique may be modified as follows.

Patient position: The patient is in the sidelying position with the spine in a rotated position.

Therapist position: The therapist stands behind the patient.

Hands: In the same position as described above.

Execution: The therapist distracts the shoulder and simultaneously provides a rotational

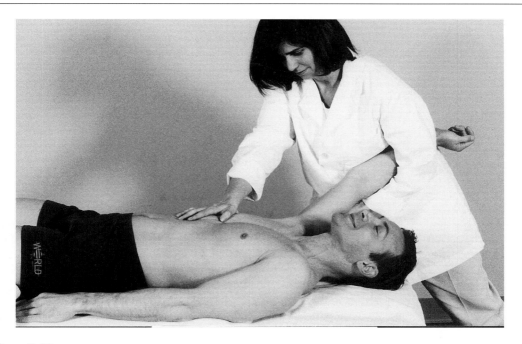

Figure 8–96

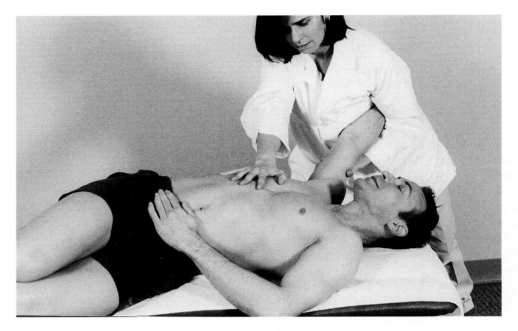

Figure 8–97

force on the spine. The other hand, which is positioned on the anterior myofascia, is moved toward the umbilicus or the contralateral ASIS. The myofascia of the anterior chest, axilla, and abdomen will be effectively stretched in this position.

Scapular Framing (Figures 8–98, 8–99, 8–100, 8–101, and 8–102)

Purpose: This technique is designed to mobilize myofascial restrictions on all three borders of the scapula. These techniques should routinely be performed on scapulothoracic problems, problems of the upper thoracic and mid-thoracic spine, cervical problems, and certain shoulder problems.

Patient position: The patient is in the sidelying position with a pillow between patient and therapist. The patient's arm should be resting comfortably on the pillow.

Therapist position: The therapist stands facing the patient with the pillow pressing against the body. There should be a "snug" fit between the patient, pillow, and therapist.

Medial Border

Hands: The top hand is lightly placed on the shoulder and the bottom hand is placed just off the medial border of the scapula, between the scapula and the thoracic spinous processes.

Execution: The shoulder is slightly retracted to slacken the tissue. As the shoulder is being retracted, the fingers of the bottom hand stroke from cephalic to caudal along the length of the medial border of the scapula.

Upper Border

Hands: The fingertips of both hands are placed over the upper trapezius muscle medially at the cervicothoracic (CT) junction. Alternately, the therapist may be at the head of the table and

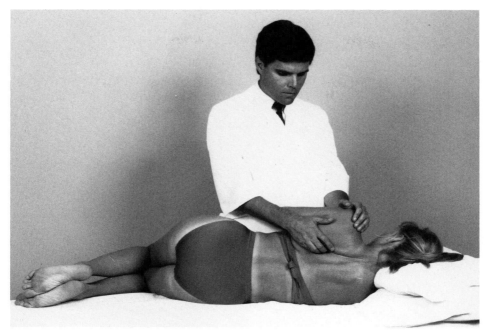

Figure 8–98

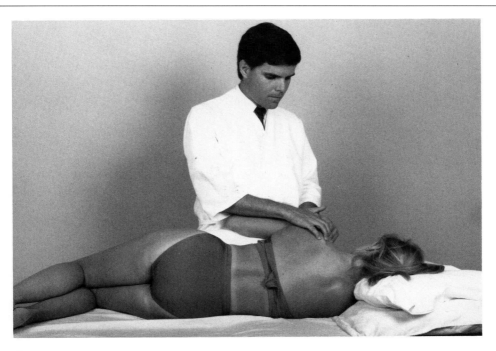

Figure 8–99

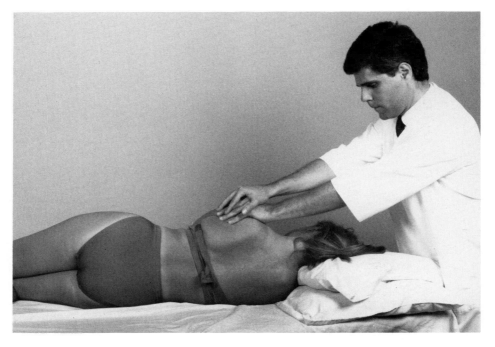

Figure 8–100

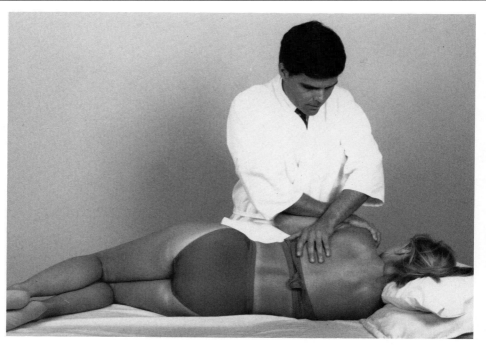

Figure 8–101

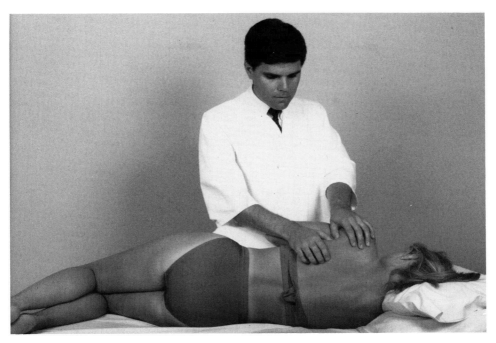

Figure 8–102

apply a caudal force, gently stretching the upper trapezius.

Execution: With firm pressure, the fingertips stroke the upper border of the scapula and upper trapezius muscle from proximal to distal (i.e., from the C/T junction to the glenohumeral joint). A gentle stretch is applied with the palms of the hand as the scapula is stroked.

Lateral Border

Hands: The palm of the bottom hand is placed over the shoulder joint to stabilize the area. The palm of the top hand is placed over the lateral border of the scapula.

Execution: With the bottom hand stabilizing the shoulder, the palm of the top hand strokes the lateral border of the scapula caudally with firm pressure. Specific finger pressure may be applied if trigger points or restrictions are found.

***Alternate Technique for Lateral Border.* Patient position:** The patient remains in the sidely-

ing position but is asked to grasp the top of the treatment table with the hand. This flexes the shoulder and tightens the myofascia in the lateral border of the scapula.

Execution: As the patient holds the treatment table, the palm of the therapist's top hand firmly strokes the lateral border of the scapula caudally. The technique may continue toward the ilium if fascial restrictions are encountered.

Scapular Mobilization (Figures 8–103 and 8–104)

Purpose: Once the scapular soft tissues have been prepared from the previously described technique, the scapula may be mobilized off the thoracic cage. This allows for more aggressive stretching of the scapulothoracic myofascia.

Patient position: The patient is in the sidelying position with a pillow between the patient and the therapist, and the patient's arm resting comfortably on the pillow.

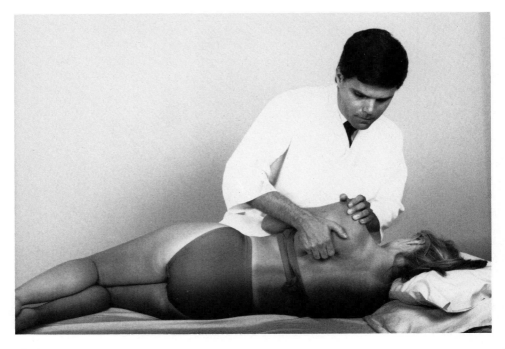

Figure 8–103

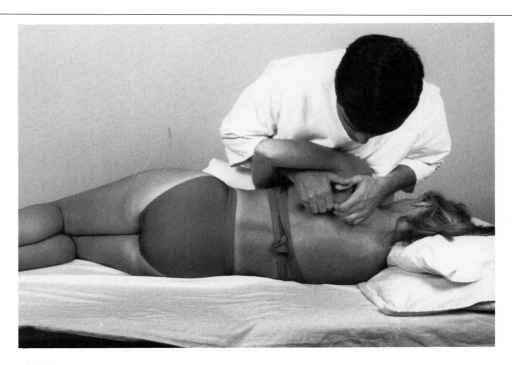

Figure 8–104

Therapist position: The therapist stands at the patient's side.

Hands: Two variations of this technique may be performed: (1) the top hand grasps the shoulder joint anteriorly. The fingers of the bottom hand slide onto the undersurface of the scapula. (2) In the alternate technique, the bottom hand slides under the arm and around the scapula until the fingers can slide onto the scapula's undersurface. The top hand also contacts the scapula so the fingers can slide onto the undersurface of the scapula. The shoulder and chest of the therapist contact the patient's shoulder anteriorly for stability.

Execution: (1) Once the fingers of the bottom hand have grasped the medial border of the scapula, the scapula and shoulder girdle complex is lifted off the thoracic cage, resulting in an aggressive stretch of the scapulothoracic myofascia. This technique succeeds if the patient is smaller than or equal in size to the therapist. (2) With both hands grasping the medial border

of the scapula, and the therapist's shoulder stabilizing anteriorly, the scapula is lifted off the thoracic cage. This technique is successful with patients who are larger in size than the therapist.

Thoracic Rotational Laminar Release (Figure 8–105)

Purpose: Previous techniques emphasize the scapulothoracic and scapulohumeral relationships and musculature. This technique penetrates to the depth of the paravertebral muscles, mobilizing the muscles and, to a certain extent the joints, into a rotational direction.

Patient position: The patient is in the sidelying position similar to the position described above.

Therapist position: Standing facing the patient with a pillow between therapist and patient.

Hands: The top hand is placed over the anterior aspect of the glenohumeral joint. The fin-

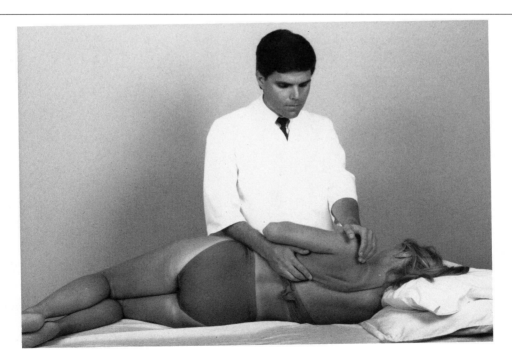

Figure 8–105

gers of the bottom hand are placed in the scapulothoracic area similar to the medial scapular framing described above.

Execution: The primary distinction between this technique and medial scapular framing is in the depth of penetration and the rotational component imparted to the thoracic spine. To execute the technique, the fingers of the bottom hand stroke cephalic to caudal with deep pressure, while the top hand is retracting the shoulder complex and rotating the thoracic spine. The fingers act as a fulcrum of rotation for the thoracic spine. If segmental restrictions are felt as the technique is being performed, the stroke may be stopped and the restricted segment may be oscillated into rotation.

Transverse Fascial Stretch of the Biceps (Figure 8–106)

Purpose: The purpose of this technique is to increase the medial/lateral mobility of the biceps in preparation for stretching or strengthening. Certain low grade peripheral entrapment neuropathies respond well when the biceps is stretched medial to lateral. This seems to free up the nerves as they pass through just posterior and medial to the biceps. Certain proximal humeral fractures cause the binding down of the biceps, and this technique will be beneficial for this type of condition as well.

Patient position: Supine.

Therapist position: The therapist will be outside the patient's arm if the treatment goes from lateral to medial, and inside the patient's arm if the technique is applied medial to lateral.

Hands: The heel of the hand is placed lateral to the muscle if the technique is going lateral to medial, and medial if the technique is going medial to lateral.

Execution: The heel of the hand pushes the biceps in a transverse direction (lateral to medial or medial to lateral) until all the "slack" is taken out of the muscle. Once the tissue is at the end

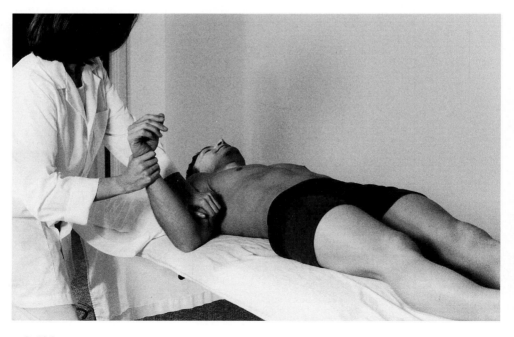

Figure 8–106

of the elastic range, the therapist pushes into the plastic range to get the final stretch. The stretch is held 3–5 seconds, then repeated.

Biceps Stretch (Figure 8–107)

Purpose: The purpose of this technique is to apply a focused stretch of the biceps muscle.

Patient position: The patient is supine with the shoulder slightly off the table. If a less aggressive version of the technique is desired, the patient may be placed in the sidelying position to accomplish a lighter version of the stretch.

Therapist position: The therapist is seated level with the patient's neck or shoulder.

Hands: The top hand is placed over the distal triceps so the fingers and thumb can wrap around the supracondylar space. The bottom hand is placed on the distal arm, just proximal to the wrist.

Execution: The therapist gently extends the patient's shoulder. At the same time the elbow is fully extended and the radioulnar joints are fully pronated. About the time the patient begins to feel a stretch, a slight traction force is placed on the arm. The therapist should ask the patient to tell when a moderate stretch is felt. Because of the long lever arm, it is difficult to tell when the biceps muscle/tendon is in a plastic stretch. After a 5–10 second hold, the arm is released and the stretch may be repeated.

Forearm "Ironing" (Figure 8–108)

Purpose: As previously described for the lumbar erector spinae, the "ironing" type techniques are useful to decrease underlying tone and move fluid. If an area is particularly tender, longitudinal stroking is always less painful than cross stroking. This technique is effective for a wide array of elbow, forearm, wrist, or hand dysfunctions. While not shown, the technique can be applied to the flexor as well as extensor surfaces of the forearm.

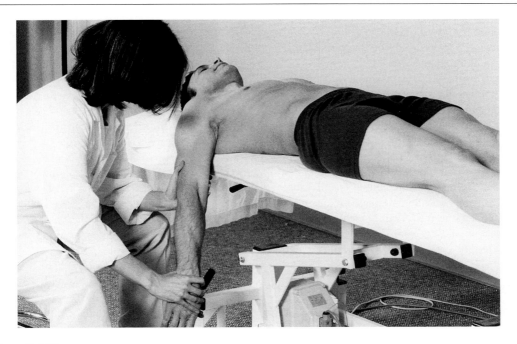

Figure 8–107

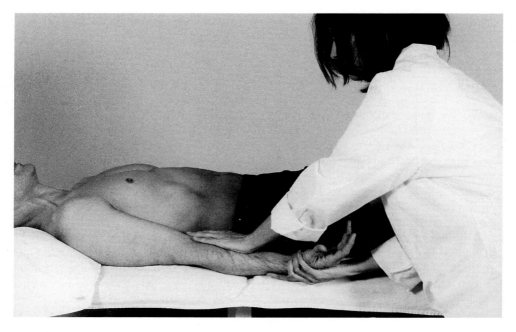

Figure 8–108

Patient position: Supine or seated, with the wrist slightly flexed (passively).

Therapist position: The therapist is positioned at the patient's side.

Hand position: The inside hand of the therapist gently grasps the wrist and flexes it. The outside hand is positioned on the distal aspect of the forearm, just proximal to the wrist.

Execution: Using a small amount of lubrication, the palm of the therapist's hand bears down on the soft tissues and begins to stroke distal to proximal, stopping at the elbow. The pressure is firm, but the hand and fingers remain relaxed, so the technique feels firm but not painful. The therapist should use some body weight to avoid the technique coming primarily from the arm.

Muscle Splay of the Forearm (Figure 8–109)

Purpose: Similar to muscle splay of the hamstring, the idea is to stroke deeply in the fascial planes separating muscles or muscle groups.

When muscle groups slide more freely on one another, their ability to be actively shortened or passively lengthened is enhanced, creating greater efficiency of contraction and/or flexibility. Treatment of the flexor surface is shown here, but the extensor surface may be treated as well.

Patient position: Supine or sitting, with the forearm on the treatment surface.

Therapist position: The therapist is positioned lateral to the patient, facing the patient.

Hands: One hand flexes the wrist, while the index and middle finger find a "wedge" between muscle groups. Alternately, the thumb can be used, but care must be taken to avoid overuse injury of the thumb.

Execution: Starting distally, the therapist wedges in between muscle groups with the index and middle finger (or thumb), applying firm pressure. Using a small amount of lubricant, the fingers slide proximally following the wedge created distally. Lack of a "wedge" or space

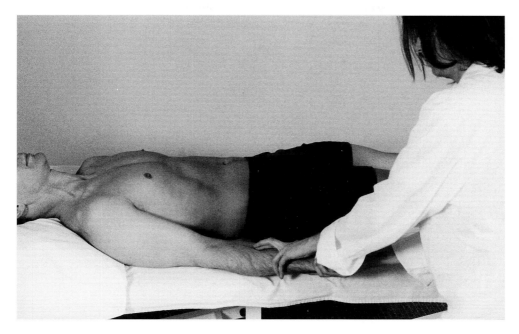

Figure 8–109

between fibers may indicate fascial adhesions. The therapist should identify and focus on these areas, working longitudinally, proximal to distal, until the fascial is freed up.

Transverse Muscle Bending of the Forearm (Figure 8–110)

Purpose: Analogous in theory to previously described muscle bending techniques, the purpose of this technique is to mobilize the forearm musculature in a transverse direction. This allows the contractile tissues to move more freely in their respective fascial compartments.

Patient position: Supine.

Therapist position: The therapist is at the patient's side using the leg to stabilize the patient's forearm.

Hands: One hand stabilizes the forearm distally. The other hand gently grasps the flexor (or extensor) surface of the forearm.

Execution: The palm of the hand pushes the muscle mass of the forearm firmly in a transverse direction through the elastic range and into the plastic range to encourage permanent deformation of the fascia. Multiple angles can be applied. For example, the flexor mass may be pushed away from or toward the ulna. The brachioradialis may be pushed anterior or posterior. The extensor surface can also be moved in either transverse direction. The therapist must "think with the hands" to determine where the restrictions are, and move in the direction of the restriction.

Palmar Stretch (Figure 8–111)

Purpose: The purpose of this technique is to stretch the palmar fascia and the palmar surface of the hand.

Patient position: Patient is supine or sitting.

Therapist position: The therapist stands facing the palm of the patient's hand.

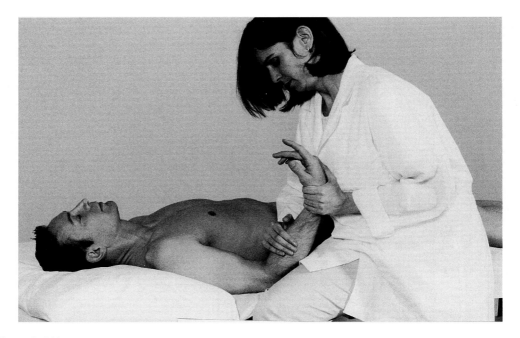

Figure 8–110

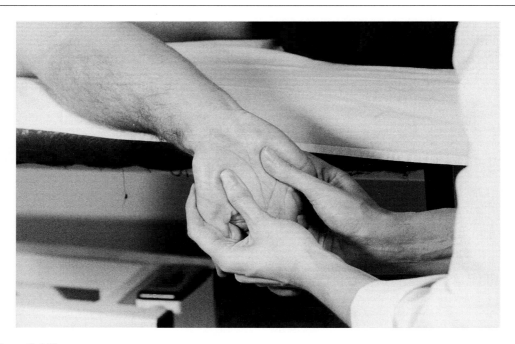

Figure 8–111

Hands: The hand position is very important in this technique. Both little fingers of the therapist are placed between the patient's index and middle fingers. The therapist's fingers are then interdigitated through the patient's fingers, with the middle and ring finger of the therapist in the web space of the patient's hand. The therapist's index fingers pull over the patient's hand, and the thumbs are available for massage during the stretch.

Execution: The therapist's fingers that are interdigitated, along with the index fingers, open the patient's hand to create a stretch. At the same time, the thumbs can be used to massage the palmar surface of the hand when the stretch is occurring. If the elbow is flexed and the wrist is in neutral, the palmar fascia will be localized. If the elbow and wrist are extended, the stretch will also include the wrist flexor muscles.

Retinacular Stretch (Figure 8–112)

Purpose: Related to the previous technique, the retinacular stretch is designed to open the carpal tunnel in a medial lateral direction, and to increase the extensibility of the retinaculum.

Patient position: Supine or sitting.

Therapist position: The therapist is facing the palmar surface of the patient's hand.

Hands: The therapist's thenar eminences are placed over the distal forearm and wrist. The fingers are on the dorsal surface of the hand to apply counter pressure.

Execution: The therapist applies firm pressure into the patient's wrist and distal forearm with the thenar eminences as the fingers apply counter pressure on the dorsal surface of the hand. A firm stretch is applied from midline outward to the ulna and radius. As the therapist's

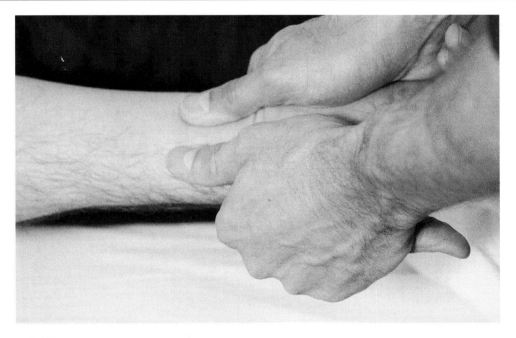

Figure 8–112

hands separate, firm pressure is maintained for maximal stretch.

TECHNIQUES FOR THE CERVICAL SPINE

Elongation of Paravertebral Muscles (Figure 8–113)

Purpose: This is a preparatory technique for other more aggressive myofascial and joint mobilization techniques. As previously defined, elongation differs from stretching in that its purpose is not necessarily to lengthen the muscle, but to elongate the spine. (Recall the analogy of elongating the accordion.) This technique, used with superficial penetration, also has a strong autonomic inhibitive effect.

Patient position: The patient lies supine with head flat on the table.

Therapist position: The therapist is seated at the head of the treatment table.

Hands: The fingers are placed over the lower cervical-upper thoracic paravertebral muscles.

Execution: The technique is executed by lightly stroking the length of the cervical paravertebral muscles from upper thoracic to subcranial. The depth of penetration may gradually be increased with progressive stroking.

Axial Flexion of the Cervical Spine (Figure 8–114)

Purpose: This technique is one of the few described in this text that can be used as either direct or indirect technique. The idea behind this indirect technique is to take the neck into the direction of restriction, thereby freeing the restriction and allowing greater axial extension. The concept is that of a dresser drawer that is stuck

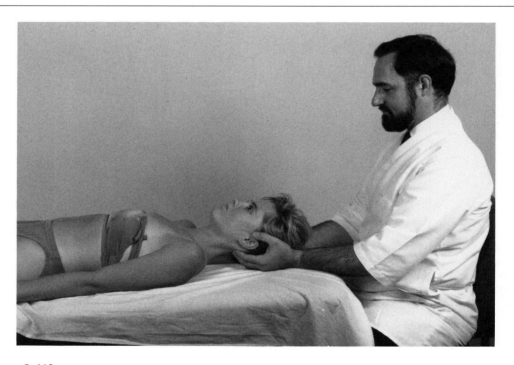

Figure 8–113

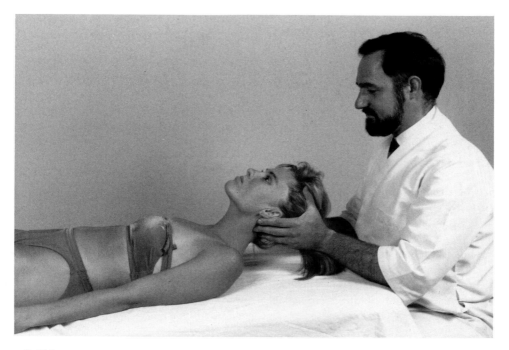

Figure 8–114

and cannot be opened. By closing the drawer, the drawer then becomes free to open. This technique can be divided into two specific components. The first is a general axial extension of the cervical spine and the second is specific axial extension at the OA joint.

Patient position: The patient is supine with the head flat on the treatment table.

Therapist position: The therapist is seated at the head of the table.

Hands: The palms of the hands cradle the base of the occiput while the fingers contact the lower cervical paravertebral musculature.

Execution: The head and neck are brought into a straight axial flexion (moving the head directly toward the ceiling). The fingers are simultaneously stroking the lower cervical paravertebrals in a medial to lateral direction. With

each repetition, the fingers are moved up a level until they are in contact with the subcranial musculature.

At this point the technique may be applied more specifically in the area of the OA joint. The head and neck are again axially flexed, with firm pressure being applied at the OA joints bilaterally with the fingertips. The fingers are no longer stroking medial to lateral, but maintaining the pressure on the OA joints. The neck may be axially extended into a diagonal plane to check for unilateral restrictions. If a unilateral OA restriction exists, the neck may be axially flexed in the same diagonal plane in an attempt to free up the restriction.

This technique may be used as a direct technique with the patients who exhibit an axially extended posture. While this posture is seen less often than the forward-head posture, the technique may be used to move the neck directly into the restriction.

Cervical Laminar Release (Figures 8–38, 8–39, and 8–115)

Sitting

Purpose: This technique is meant to elongate the cervical paravertebral musculature and to improve cervical forward bending.

Patient position: Sitting.

Therapist position: The therapist is standing behind the patient.

Hands: In the bilateral technique, both hands are placed on the paravertebral muscles with the thumbs and PIP of the index finger contacting the patient. In the unilateral technique, one hand is on the patient's head to monitor the diagonal movement of the patient's head and neck.

Execution: The patient is first asked to forward bend the cervical spine segmentally. As the flexion occurs, the hands stroke caudally through the midcervical, cervicothoracic, and upper thoracic areas. If unilateral technique is preferred, the monitoring hand gently guides the patient into a diagonal pattern as the other hand gently strokes unilaterally through the cervical, cervicothoracic, and upper thoracic areas.

Supine

Purpose: Elongation of the cervical myofascia.

Patient position: Supine.

Therapist position: Seated at the head of the table.

Hands: One hand cradles the head at the occiput and brings the cervical spine into a forward-bent position. The other hand makes contact with the cervical paravertebral muscles,

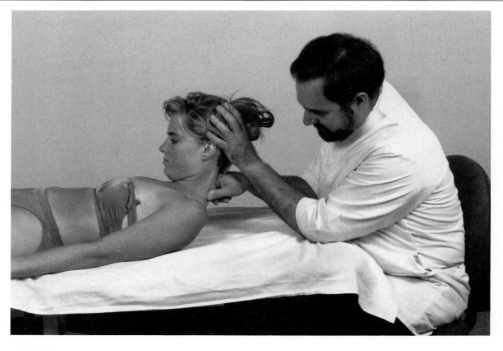

Figure 8–115

bilaterally, using the thumb on one side, and the PIP of the index finger on the other side.

Execution: One hand holds the neck statically in the forward-bent position while the other hand strokes gently from approximately midcervical to cervicothoracic junction.

Diagonal Stretch of Cervical Cervicothoracic Myofascia (Figure 8–116)

Purpose: This technique stretches the posterior myofascial structures as well as the upper trapezius and levator scapula muscles.

Patient position: Supine.

Therapist position: Seated at the head of the table.

Hands: One hand cradles and positions the head in a combination of forward bending, side bending, and rotation. The rotation can be to either the same or the opposite side as the forward bending depending on the restriction. The

other hand is placed firmly on the patient's shoulder.

Execution: With the patient positioned, gentle to moderate pressure is applied caudally on the shoulder while a pressure is applied with the other hand into forward bending, side bending, and rotation.

Manipulation of Subcranial and 0A Myofascia (Figure 8–117)

Purpose: This technique is useful in releasing subcranial myofascia as well as for mobilizing the 0A joints. This technique allows patient participation and, as such, may be considered a muscle energy technique. The idea behind the technique is stabilization of the occiput and movement of the atlas. The patient is axially flexing and extending the neck while the occiput is held rigid.

Patient position: Supine.

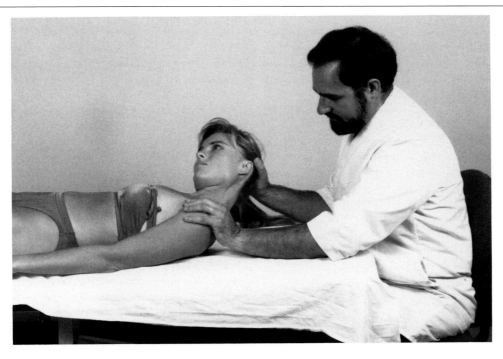

Figure 8–116

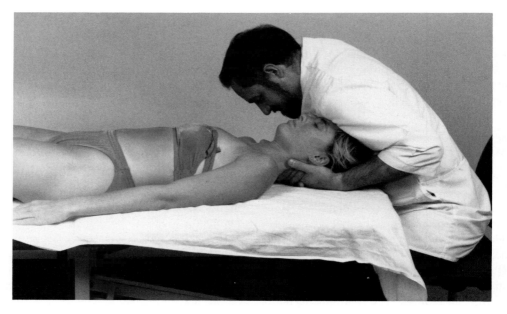

Figure 8–117

Therapist position: The therapist may be standing or sitting. The patient's head will be cradled by the therapist's arm and shoulder.

Hands: As the patient's head is cradled with one arm and shoulder of the therapist, the hand firmly grasps the occiput. The other hand is placed over the hand grasping the occiput as additional reinforcement.

Execution: With the therapist firmly holding the head, the patient is asked gently to axially flex and extend the neck. The head is not allowed to move, so the neck is actually moving on the head. The atlas is allowed to translate anteriorly and posteriorly on a nonmoving occiput. After several repetitions, the patient is allowed to rest his or her head on the table and the amount of resting axial flexion is reassessed.

Masseters-TMJ Decompression (Figure 8–118)

Purpose: Prior to any intraoral soft tissue manipulation, the clinician should always at-

tempt extraoral soft tissue manipulation in restoring mobility of the temporomandibular joint (TMJ). This technique inhibits the masseters, allowing for a more comfortable and increasingly functional opening of the mandible. The functional opening may be significantly increased without having to perform intraoral maneuvers.

Patient position: The patient is supine with the head flat on the treatment table.

Therapist position: The therapist is seated at the head of the table.

Hands: The tips of the index, middle, and ring fingers are placed on the masseters just below the temporomandibular joint line.

Execution: With moderate depth of pressure, the therapist strokes along the length of the masseters away from the TMJ. After several strokes, the patient is asked to open the mouth in a subtle and relaxed manner as the stroke is being applied. As the masseters are stroked, the relaxed mandible will open further and a gentle opening stretch may be applied at the end of the technique.

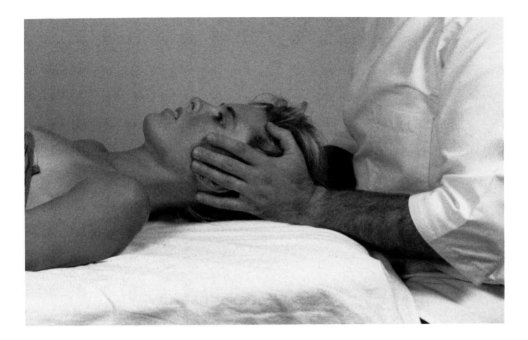

Figure 8–118

Frontal Facial Decompression (Figure 8–119)

Purpose: The purpose of this technique is twofold. First, the hand position can be used to provide a gentle subcranial traction. Second, the technique can be used to inhibit the frontalis muscle or to provide a fascial stretch to the frontal, nasal, and facial fascias. This is useful in cases of parieto-occipital headaches or sinus headaches.

Patient Position: Supine.

Therapist position: The therapist is seated at the head of the table.

Hands: One hand gently cradles the occiput, while the other hand is placed directly over the frontal area of the patient's face, with the therapist's thumb pointing in the direction of the therapist.

Execution: The therapist gives a slight traction with the bottom hand. With the palmar surface of the top hand in full contact over the fron-

talis, a fascial traction is simultaneously applied, and held for 15–30 seconds. The emphasis of this technique is on the frontal fascial stretch and frontal decompression.

Retro-Orbital Decompression (Figure 8–120)

Purpose: Related to the previous technique, the purpose is to stretch the retro-orbital fascia and the fascia around the nasal suture. This technique is especially indicated for patients with retro-orbital headaches and sinus headaches.

Patient position: Supine.

Therapist position: Seated at the head of the table.

Hands: The bottom gently cradles the base of the occiput. The palm of the top hand makes contact with the frontal area, while the fingers are positioned as follows: The index and ring finger are placed over the left and right orbital

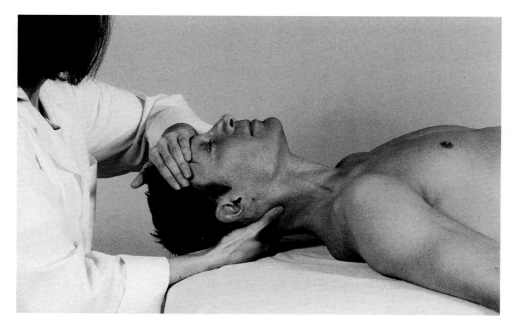

Figure 8–119

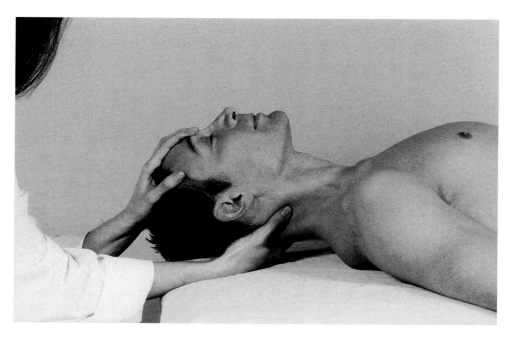

Figure 8–120

bones, just inside the eyebrow, well away from the eyes. The middle finger is placed just over the nasal suture.

Execution: A gentle traction is applied through the occiput with the bottom hand. The palm of the top hand places a mild traction over the frontal fascia, while the fingers apply a fascial traction over the retro-orbital and nasal fasciae. Care must be taken to make absolutely no contact with the eyes. The fascial stretch is applied firmly with the pads of the fingers for 10–20 seconds.

Sternocleidomastoids (Figures 8–121 and 8–122)

Purpose: This technique decreases tone of the sternocleidomastoid (SCM) muscles. Even if the muscle is relaxed in the supine position, the

SCM may still be exquisitely tender to palpation due to overuse in the erect posture.

Patient position: The patient is supine with the head off the edge of the table.

Therapist position: The therapist is seated at the head of the table, gently cradling the patient's head in a very slight backward bent position.

Hands: One hand is cradling the occiput, while the other hand is positioned with the thumb placed on the cephalic portion of the SCM near the mastoid process.

Execution: The therapist rotates the patient's neck and adds a slight amount of backward bending of the cervical spine. The thumb of the other hand is placed on the SCM near the insertion at the mastoid process. The SCM is gently stroked from cephalic to caudal. The SCM may also be cross-stroked at any point along the muscle belly where trigger points, tender areas, or areas of hypertonicity are encountered.

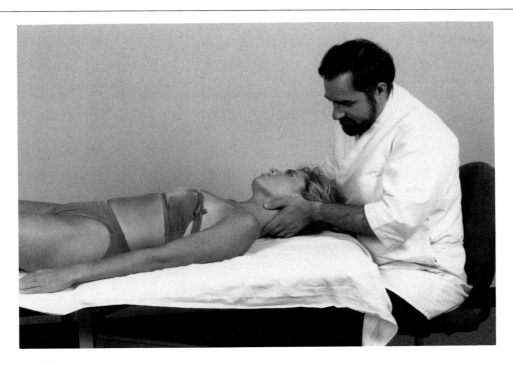

Figure 8–121

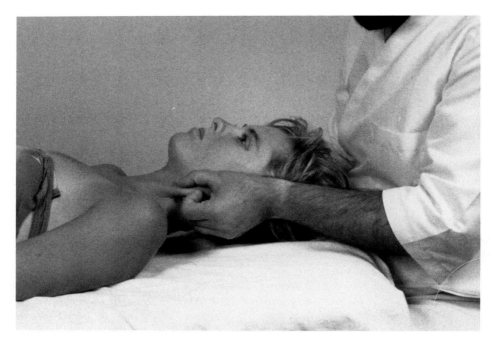

Figure 8–122

REFERENCES

1. Feldenkrais M. *Advances through Movement*. New York: Harper & Row, 1972.
2. Rosenthal E. The Alexander technique—What it is and how it works. *Medical Problems of Performing Artists*, 1987 (June): 53–57.
3. Dietze E, Schliack H, et al. *A Manual of Reflexive Therapy of the Connective Tissues*. Scarsdale, NY: Sidney Simon, 1978.

Index